The
PHYSICS
Devotional

CELEBRATING *the* WISDOM *and* BEAUTY *of* PHYSICS

Clifford A. Pickover

STERLING
New York

STERLING

New York

An Imprint of Sterling Publishing
1166 Avenue of the Americas
New York, NY 10036

ISBN 978-1-4549-1554-6

Distributed in Canada by Sterling Publishing
℅ Canadian Manda Group, 664 Annette Street
Toronto, Ontario, Canada M6S 2C8
Distributed in the United Kingdom by GMC Distribution Services
Castle Place, 166 High Street, Lewes, East Sussex, England BN7 1XU
Distributed in Australia by Capricorn Link (Australia) Pty. Ltd.
P.O. Box 704, Windsor, NSW 2756, Australia

For information about custom editions, special sales, and premium and corporate
purchases, please contact Sterling Special Sales at 800-805-5489
or specialsales@sterlingpublishing.com.

Manufactured in China

2 4 6 8 10 9 7 5 3 1

www.sterlingpublishing.com

CONTENTS

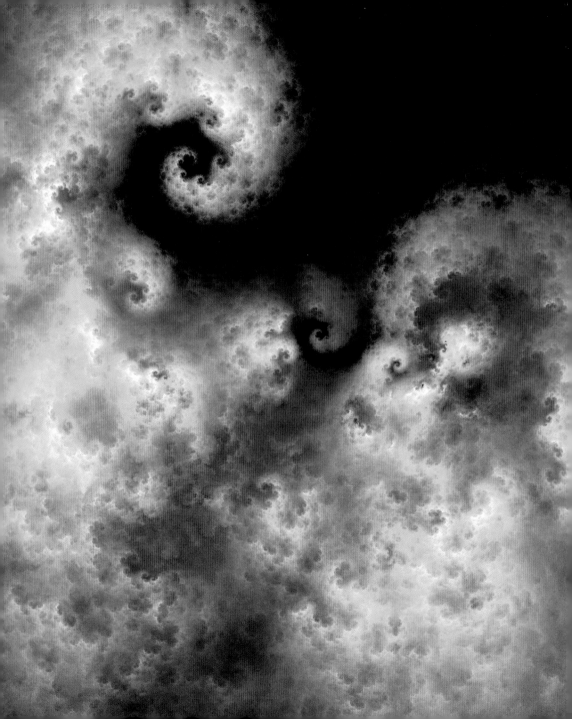

INTRODUCTION

A Universe of Laws

Today, physicists roam far and wide, studying an awesome variety of topics, size scales, and fundamental laws in order to understand the behavior of nature, the universe, and the very fabric of reality. Physicists ponder multiple dimensions, parallel universes, and the possibilities of wormholes connecting different regions of space and time. The discoveries of physicists often lead to new technologies and even change our philosophies and the way we look at the world. For example, for many scientists, the Heisenberg uncertainty principle means that the physical universe literally does not exist in a determinist form but is considered as a collection of probabilities. Of course, physics can also be eminently practical. Advances in the understanding of electromagnetism led to the invention of the radio, television, and computers. Likewise, an understanding of thermodynamics helped lead to more efficient cars.

As will become apparent as you peruse this book, the precise scope of physics has not been fixed through the ages, nor is it easily delimited. I have taken a rather wide view and have included topics that touch on engineering and applied physics, advances in our understanding of the nature of astronomical objects, and even a few topics that are quite philosophical. Despite this large scope, most areas of physics have in common a strong reliance on mathematical tools to aid scientists in their understandings, experiments, and predictions about the natural world.

You'll note that the quotations in this book come from diverse sources, and though many are from important physicists, I have intentionally tried to include a range of sources—from serious scientists and educators to novelists like John Steinbeck, Douglas Adams, Larry Niven, and Robert Heinlein. Even diverse thinkers such as Edgar Allan Poe and Vincent van Gogh make brief appearances.

You'll also discover that a significant number of the quotations concern physicists' views on the laws of nature, mysticism, and even religion. Centuries ago, many physicists saw God's hand in nature's laws. For example, British scientist Sir Isaac Newton (1642–1727), one of the most influential scientists in all of history, and many of his contemporaries believed that the laws of the universe were established by the will of God, who acted in a logical manner. The Irish natural philosopher Robert Boyle (1627–1691) was quite devout and loved the Bible. His constant desire to understand God drove his interest in discovering laws of nature.

This book also includes a number of quotations regarding Isaac Newton. Patricia Fara, author of *Newton: The Making of Genius*, wrote, "Even the briefest survey of Newton's life unsettles his image as the idealized prototype of a modern scientist. . . . A renowned expert on Jason's fleece, Pythagorean harmonics, and Solomon's temple, his advice was also sought on the manufacture of coins and remedies for headaches. . . . Newton had no laboratory team to supervise . . . and never travelled outside eastern England." John Maynard Keynes, in "Essays in Biography: Newton, the Man," wrote, "Newton was not the first of the age of reason. He was the last of the magicians, the last of the Babylonians and Sumerians, the last great mind which looked out on the world with the same eyes as those who began to build our intellectual inheritance rather less than ten thousand years ago."

In the pages that follow, I have also included a variety of subjects that may exist at the edges of physics and its fractal boundaries with science fiction—including topics on parallel worlds, higher dimensions, black holes and time travel (two particularly great interests of mine), string theory, and the mysterious nature of ultimate reality. Quantum mechanics gives us a glimpse of a world that is so strangely counterintuitive that it raises questions about space, time, information, and cause and effect. Moreover, science fiction has often been a useful source of scientific ideas. I like to think of science fiction as the "literature of edges" because the topics are poised on the edge of what is and what might be. Certainly, science fiction is a literature of change. Additionally, our universe is a science-fiction universe, filled with puzzles to solve—constantly fluctuating and evolving. Isaac Asimov suggested that science fiction is the only form of literature that consistently considers the nature of the changes that face us, along with possible solutions.

This devotional also offers a way of providing glimpses of physics and aesthetics—presenting wisdom and poetry in brief, hopefully inspiring readers to learn more about the universe of physics. I hope there are future editions of this book, and welcome suggestions and corrections from readers.

Of course, although this book is not a "daily devotional" in the traditional religious sense of the phrase, I hope that contemplating the quotations and images in this book fills your mind with wonder and astonishment—while stretching your imagination and serving as a source of beauty. For example, reading a quotation a day, before going to work or school, or before sleep, seems to place our ordinary life tasks within a wider perspective. Perhaps some theoretical physicists, like priests or theologians, have been seeking "ideal," immutable, nonmaterial truths and then venturing to apply these truths to the real world. They seek to understand the patterns and order in the cosmos, and crave inspiration.

Physics can be among the most difficult of sciences. Our physics-centered description of the universe grows forever, but our brains and language skills remain entrenched. New kinds of physics are uncovered through time, but we need fresh ways to think and to understand. Who knows what the future of physics will offer? Toward the end of the nineteenth century, prominent physicist William Thomson, also known as Lord Kelvin, proclaimed the end of physics. He could never have foreseen the rise of quantum mechanics and relativity—and the dramatic changes these areas would have on the field of physics. In the early 1930s physicist Ernest Rutherford said of atomic energy, "Anyone who expects a source of power from the transformation of these atoms is talking moonshine." In short, predicting the future of the ideas and applications of physics is difficult, if not impossible.

Discoveries in physics provide a framework in which to explore the subatomic and supergalactic realms, and the concepts of physics allow scientists to make predictions about the universe. It is a field in which philosophical speculation can provide a stimulus for scientific breakthroughs. The discoveries in physics are among humanity's greatest achievements. For me, physics cultivates a perpetual state of wonder about the limits of thought, the workings of the universe, and our place in the vast space-time landscape we call home.

Physicist Micro-Biographies

The birthdays of some famous and important physicists are highlighted throughout this book with "Born on This Day" designations. You can turn to the back of the book for micro-biographies that provide just a taste of the advanced fields these unique individuals explored, along with the countries with which they are associated, and a few curious facts that interest me personally about these individuals.

JANUARY 1

"We all use physics every day.
When we look in a mirror or put on a pair of glasses we are
using the physics of optics. When we set our alarm clocks we track time;
when we follow a map we navigate geometric space.
Our cell phones connect us via invisible electromagnetic threads
to satellites orbiting overhead. But physics is not all about technology. . . .
Even the blood flowing through our arteries follows laws of physics,
the science of the physical world."

—JOANNE BAKER, *50 PHYSICS IDEAS YOU REALLY NEED TO KNOW*, 2007

JANUARY 2

BORN ON THIS DAY: Rudolf Clausius, 1822

"Clausius's Law of Entropy Nonconservation was like saying that a casino's
positive money changes always exceeded its negative money changes.
In other words, a casino's winnings always exceeded its losses;
it always made a profit, which was how it stayed in business.
A casino existed at the expense of its players, which meant it could
keep winning only so long as its players could keep losing.
When they had lost everything, when positive money changes ceased to exist,
the casino would shut down forever."

—MICHAEL GUILLEN, *FIVE EQUATIONS THAT CHANGED THE WORLD*, 1995

JANUARY 3

"A good many times I have been present at gatherings of people who, by the standards of the traditional culture, are thought highly educated and who have with considerable gusto been expressing their incredulity at the illiteracy of scientists. Once or twice I have been provoked and have asked the company how many of them could describe the Second Law of Thermodynamics. The response was cold: it was also negative. Yet I was asking something which is the scientific equivalent of: *Have you read a work of Shakespeare's?*"

—C. P. SNOW, "THE TWO CULTURES," REDE LECTURE, 1959

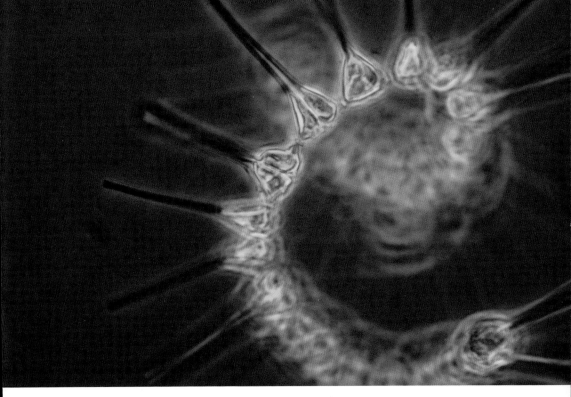

JANUARY 4

"Man is . . . related inextricably to all reality, known and unknowable. . . .
All things are one thing and that one thing is all things—plankton,
a shimmering phosphorescence on the sea and the spinning planets and an
expanding universe, all bound together by the elastic string of time.
It is advisable to look from the tide pool to the stars
and then back to the tide pool again."

—JOHN STEINBECK, *THE LOG FROM THE SEA OF CORTEZ*, 1951

$$F_g = G \frac{m_1 m_2}{r^2}$$

JANUARY 5

"There is a kind of symbiotic relationship here between law and theory.
A theory becomes more and more respected and powerful
the more phenomena that can be derived from it, and the law describing these
phenomena becomes more meaningful and useful if it can be made part of a theory.
Thus, Newton's theory of universal gravitation gained greatly in stature
because it enabled one to derive the laws that govern the moon's motion,
known by empirical rules since the days of the Babylonian observers."

—ARNOLD ARONS IN *DEVELOPMENT OF CONCEPTS OF PHYSICS*, 1965

JANUARY 6

"Consider the true picture.
Think of myriads of tiny bubbles, very sparsely scattered,
rising through a vast black sea. We rule some of the bubbles.
Of the waters we know nothing. . . ."

—LARRY NIVEN AND JERRY POURNELLE, *THE MOTE IN GOD'S EYE*, 1993

JANUARY 7

"There is no philosophy which is not founded upon knowledge
of the phenomena, but to get any profit from this knowledge it is absolutely
necessary to be a mathematician."

—DANIEL BERNOULLI, LETTER TO JOHN BERNOULLI III, JANUARY 7, 1763

JANUARY 8

"Black holes are the bane of modern astronomy.
Invisible by definition, their existence has proved difficult to substantiate,
yet black holes have captured the popular imagination
in a way that no other astronomical objects have succeeded in doing.
Black holes are time machines, as well as openings to other universes."

—JOSEPH SILK, FOREWORD TO JEAN-PIERRE LUMINET'S *BLACK HOLES*, 1992

JANUARY 9

"I became possessed with the keenest curiosity about the whirl itself.
I positively felt a wish to explore its depths, even at the sacrifice I was going to make;
and my principal grief was that I should never be able to tell
my old companions on shore about the mysteries I should see."

—EDGAR ALLAN POE, "A DESCENT INTO THE MAELSTRÖM," 1841

JANUARY 10

"Time is a relationship that we have with the universe;
or more accurately, we are one of the clocks, measuring one kind of time.
Animals and aliens may measure it differently. We may even be able to change
our way of marking time one day, and open up new realms of experience,
in which a day today will be a million years."

—GEORGE ZEBROWSKI, "TIME IS NOTHING BUT A CLOCK," *OMNI*, 1994

JANUARY 11

"Space isn't remote at all. It's only an hour's drive away
if your car could go straight upwards."

—FRED HOYLE, "SAYINGS OF THE WEEK," *THE OBSERVER*, 1979

JANUARY 12

"A correct theory is one that can presumably be verified by experiment. And yet, in some cases, scientific intuition can be so accurate that a theory is convincing even before the relevant experiments are performed. Einstein— and many other physicists as well—remained convinced of the truth of special relativity even when . . . experiments seemed to contradict it."

—RICHARD MORRIS, *DISMANTLING THE UNIVERSE*, 1984

JANUARY 13

"Even if the rules of nature are finite, like those of chess,
might not science still prove to be an infinitely rich, rewarding game?"

—JOHN HORGAN, "THE NEW CHALLENGES," *SCIENTIFIC AMERICAN*, 1992

JANUARY 14

"The mathematical take-over of physics has its dangers,
as it could tempt us into realms of thought which embody mathematical
perfection but might be far removed, or even alien to, physical reality.
Even at these dizzying heights we must ponder the same
deep questions that troubled both Plato and Immanuel Kant.
What is reality? Does it lie in our mind,
expressed by mathematical formulae, or is it 'out there'?"

—MICHAEL ATIYAH, "PULLING THE STRINGS," *NATURE*, 2005

JANUARY 15

"It is known that there are an infinite number of worlds,
simply because there is an infinite amount of space for them to be in.
However, not every one of them is inhabited. Therefore, there must be
a finite number of inhabited worlds. Any finite number divided by infinity
is as near to nothing as makes no odds, so the average population
of all the planets in the Universe can be said to be zero.
From this it follows that the population of the whole Universe is also zero,
and that any people you may meet from time to time are merely
the products of a deranged imagination."

—DOUGLAS ADAMS, *THE RESTAURANT AT THE END OF THE UNIVERSE*, 1980

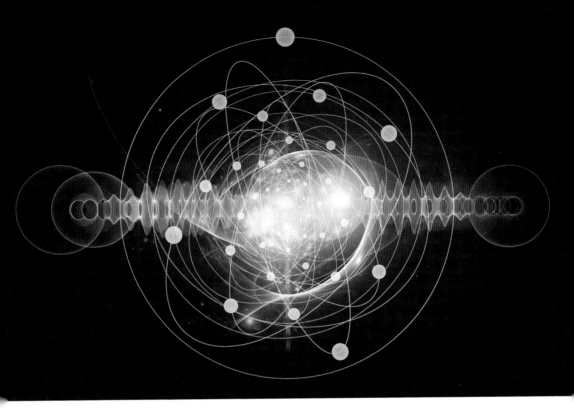

JANUARY 16

"[Quantum mechanics] is absolutely central to explaining why
atoms did not collapse, how solids can be rigid, and how different atoms combine
together in what we call chemistry and biology. . . .
But this triumph of quantum mechanics came with an unexpected problem—
when you stepped outside of the mathematics and tried to explain
what was going on, it didn't seem to make any sense."

—PAUL QUINCEY, "WHY QUANTUM MECHANICS IS NOT SO WEIRD AFTER ALL,"
SKEPTICAL INQUIRER, 2006

JANUARY 17

"During the Renaissance . . . the interaction among different
European cultures stimulated creativity through new ways of thinking
and new paradigms for the observation of nature. . . .
The foundation of scientific academies, notably the Accademia dei Lincei,
the Royal Society, and the Académie des Sciences, and the establishment
of universities throughout Western Europe, contributed to scientific progress. . . ."

—MAURIZIO IACCARINO, "SCIENCE AND CULTURE," *EMBO REPORTS*, 2003

JANUARY 18

"Science is an accumulation of written narratives about our relationship to nature."

—JOSEPH SCHWARTZ, *THE CREATIVE MOMENT:*
HOW SCIENCE MADE ITSELF ALIEN TO MODERN CULTURE, 1992

JANUARY 19

"Physics is crucial to understanding the world around us, the world inside us, and the world beyond us. It is the most basic and fundamental science. Physics challenges our imaginations with concepts like relativity and string theory, and it leads to great discoveries, like computers and lasers, that lead to technologies which change our lives. . . . Physics encompasses the study of the universe from the largest galaxies to the smallest subatomic particles. Moreover, it's the basis of many other sciences, including chemistry, oceanography, seismology, and astronomy. . . ."

—AMERICAN PHYSICAL SOCIETY, "WHY STUDY PHYSICS?," WWW.APS.ORG

JANUARY 20

BORN ON THIS DAY: André-Marie Ampère, 1775

"Few men of the nineteenth century are so interesting as André Marie Ampère, who is . . . deservedly spoken of as the founder of the science of electro-dynamics. Extremely precocious as a boy . . . he grew up, indeed, to be a young man of the widest possible interests. . . . [Dominique] Arago . . . has said of Ampère's discovery identifying magnetism and electricity that 'the vast field of physical science perhaps never presented so brilliant a discovery, conceived, verified, and completed with such rapidity.'"

—MICHAEL O'REILLY AND JAMES WALSH, *MAKERS OF ELECTRICITY*, 1909

JANUARY 21

"This present universe has evolved from an unspeakably unfamiliar earlier condition,
and faces a future extinction of endless cold or intolerable heat.
The more the universe seems comprehensible, the more it also seems pointless. . . .
The effort to understand the universe is one of the very few things that lifts
human life a little above the level of farce, and gives it some of the grace of tragedy."

—STEVEN WEINBERG, *THE FIRST THREE MINUTES*, 1977

JANUARY 22

"When it comes to the world of the quantum, we really are crossing into a quite extraordinary domain . . . where it seems we are free to choose any one of a number of explanations for what is observed, each of which is in its way so astonishingly strange that it even makes tales of alien abductions sound perfectly reasonable."

—JIM AL-KHALILI, *QUANTUM: A GUIDE FOR THE PERPLEXED*, 2004

JANUARY 23

"Archimedes will be remembered when Aeschylus is forgotten, because languages die and mathematical ideas do not. 'Immortality' may be a silly word, but probably a mathematician has the best chance of whatever it may mean."

—G. H. HARDY, *A MATHEMATICIAN'S APOLOGY*, 1941

JANUARY 24

"Science advances but slowly, with halting steps.
But does not therein lie her eternal fascination? And would we not soon tire
of her if she were to reveal her ultimate truths too easily?"

—KARL VON FRISCH, *A BIOLOGIST REMEMBERS*, 1967

JANUARY 25

"The earliest use of the term [law of nature] in English . . . dates only from the seventeenth century, when systematic science began to take off. The first two examples traced by the *Oxford English Dictionary* are dated 1665—one from the Transactions of the Royal Society and one from Boyle—and they relate to a universe set and maintained in motion by the command of God. . . . Descartes presents in the *Principia Philosophiae* (1644) . . . certain rules or laws of nature."

—MICHAEL FRAYN, *THE HUMAN TOUCH*, 2007

JANUARY 26

"The art of conversing with stones is called physics.
The question-and-answer periods of the conversations are called experiments.
The usual talk is about sizes, temperatures, densities, motion, causes and effects,
and the nature of physical space and time.
The language spoken is mathematics."

—JULIUS THOMAS FRASER, *TIME, THE FAMILIAR STRANGER*, 1987

JANUARY 27

"Mathematical physics represents the purest image that the view of nature may generate in the human mind; this image presents all the character of the product of art; it begets some unity, it is true and has the quality of sublimity; this image is to physical nature what music is to the thousand noises of which the air is full . . ."

—THÉOPHILE DE DONDER, AS QUOTED BY ILYA PRIGOGINE, NOBEL PRIZE LECTURE, 1977

JANUARY 28

"The search for FTL [faster-than-light] communicators in the cracks
of present-day physics has been compared to the nineteenth-century search
for perpetual-motion machines. In trying to understand clearly why perpetual-motion
machines invariably failed to work, physicists were led to the formulation
of the first and second laws of thermodynamics which govern the
amount and quality of energy available in any conceivable physical system.
In a like manner, the study of why certain FTL schemes fail may also lead to certain
general laws which on the surface seem to have nothing at all to do with the
achievement of high-velocity communication."

—NICK HERBERT, *FASTER THAN LIGHT*, 1988

JANUARY 29

BORN ON THIS DAY: Mohammad Abdus Salam, 1926

"The advantage to being a theoretical physicist, [Yakir] Aharonov says, is that you never have to worry about the cost of a thought experiment."

— DAVID FREEDMAN, "TIME TRAVEL REDUX," *DISCOVER*, 1992

JANUARY 30

"We could imagine a world in which
causality does not lead to a consistent order of *earlier* and *later*.
In such a world the past and the future would not be irrevocably separated,
but could come together in the same present, and we could meet our former selves
of several years ago and talk to them. It is an empirical fact that our world
is not of this type. . . . Time order reflects the causal order of the universe."

—HANS REICHENBACH, *THE RISE OF SCIENTIFIC PHILOSOPHY*, 1951

JANUARY 31

"In all the history of mankind, there will be only one generation
that will be first to explore the solar system, one generation for which, in childhood,
the planets are distant and indistinct discs moving through the night sky,
and for which, in old age, the planets are places, diverse new worlds
in the course of exploration."

—CARL SAGAN, *THE COSMIC CONNECTION*, 1973

FEBRUARY 1

"Our physical science does not necessarily deal with reality, whatever that is. Rather, it has merely generated a set of consistency relationships to explain our common ground of experience. . . . We have developed these mathematical laws based ultimately on a set of definitions of mass, charge, space, and time. We don't really know what these quantities are, but we have defined them to have certain unchanging properties and have thus constructed our edifice of knowledge on these pillars."

—WILLIAM A. TILLER, PREFACE TO ITZHAK BENTOV'S *STALKING THE WILD PENDULUM*, 1976

FEBRUARY 2

"When Newton worked out the force of gravity, he helped to set into motion the industrial revolution. When Faraday worked out electricity and magnetism, he set into motion the electric age. When Einstein wrote down $E = mc^2$, he unleashed the nuclear age. Now, we are on the verge of a theory of all forces which may, one day, determine the fate of the human species."

—MICHIO KAKU, "BBC INTERVIEW ON PARALLEL UNIVERSES," WWW.BBC.CO.UK, 2002

FEBRUARY 3

"We are like the explorers of a great continent who have penetrated its margins
in most points of the compass and have mapped the major mountain chains and rivers.
There are still innumerable details to fill in, but the endless horizons no longer exist."

—H. BENTLEY GLASS, IN A SPEECH
TO THE AMERICAN ASSOCIATION FOR THE ADVANCEMENT OF SCIENCE, 1970

FEBRUARY 4

"Good theories [don't necessarily] convey ultimate truth, or [imply]
that there 'really are' little hard particles rattling around against each other
inside the atom. Such truth as there is in any of this work lies in the mathematics;
the particle concept is simply a crutch ordinary mortals can use to help
them towards an understanding of the mathematical laws."

—JOHN GRIBBIN, *THE SEARCH OF SUPERSTRINGS, SYMMETRY, AND THE THEORY OF EVERYTHING*, 2000

FEBRUARY 5

"Though my soul may set in darkness, it will rise in perfect light;
I have loved the stars too fondly to be fearful of the night."

—SARAH WILLIAMS, "THE OLD ASTRONOMER TO HIS PUPIL," 1868

FEBRUARY 6

"I take the positivist viewpoint that a physical theory is just a mathematical model and that it is meaningless to ask whether it corresponds to reality. All that one can ask is that its predictions should be in agreement with observation. . . . I don't demand that a theory correspond to reality because I don't know what [reality] is. . . . All I'm concerned with is that the theory should predict the results of measurements."

—STEPHEN HAWKING, *THE NATURE OF SPACE AND TIME*, 1996

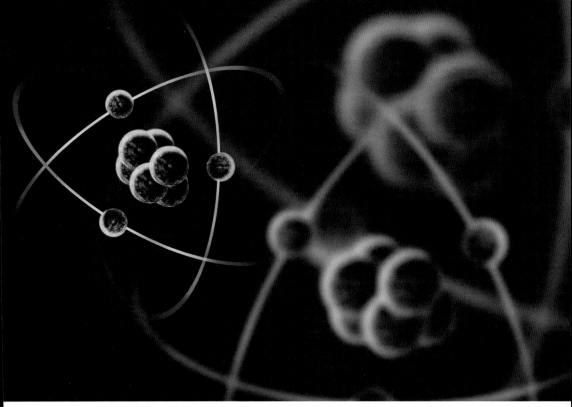

FEBRUARY 7

"Nobel prize-winning physicist Max Born . . . boasted to some visitors
at Göttingen University back in 1928 that 'physics, as we know it,
will be over in six months.' In other words, there was a unity in modern cosmology,
a belief that everything would fall into place as an all-embracing system
that left behind no puzzles."

—EDMUND E. JACOBITTI, *COMPOSING USEFUL PASTS*, 2000

FEBRUARY 8

"It is the nature of engineering to take a natural, often subtle effect and control it,
with a view toward greatly leveraging and magnifying it. . . . Consider, for example,
how we have focused and amplified the subtle properties of Bernoulli's principle
(that air rushing over a curved surface has a slightly lower pressure
than it does over a flat surface) to create the whole world of aviation."

—RAY KURZWEIL, IN JOHN BROCKMAN'S *WHAT WE BELIEVE BUT CANNOT PROVE*, 2006

FEBRUARY 9

"While every one of us is a time traveler,
the cosmic pathos that elevates human history to the level of tragedy arises precisely
because we seem doomed to travel in only one direction—into the future."

—LAWRENCE M. KRAUSS, *THE PHYSICS OF STAR TREK*, 2007

FEBRUARY 10

"One criterion for declaring the program of mechanics to be successful
would be the discovery that simple laws . . . do indeed exist.
This turns out to be the case, and this fact constitutes the essential reason
that we 'believe' the laws of classical mechanics. If the force laws
had turned out to be very complicated, we would not be left
with the feeling that we had gained much insight into the workings of nature."

—DAVID HALLIDAY AND ROBERT RESNICK, *PHYSICS*, 1966

FEBRUARY 11

BORN ON THIS DAY: Josiah Willard Gibbs, 1839

"Reality might not behave always as we've come to expect.
It's H. P. Lovecraft's crawling chaos, Nyarlathotep; it's the heart of a Stephen Hawking
black hole; it's the core of Benoit Mandelbrot's fractals; in fact,
it's Albert Einstein's assertion that, after all, physics as we know it
might well be a local phenomenon."

—EUGENE R. STEWART, "SHADES OF MEANING:
SCIENCE FICTION AS A NEW METRIC," *SKEPTICAL INQUIRER*, 1996

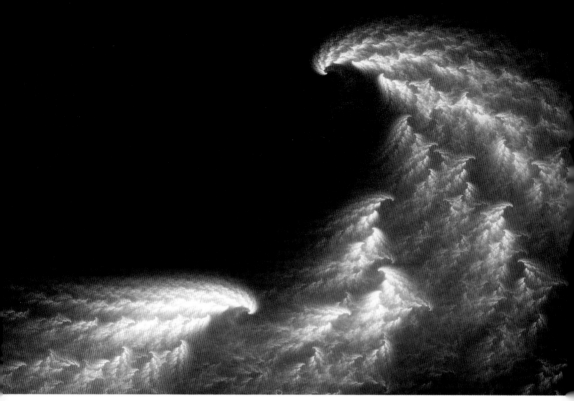

FEBRUARY 12

"Perhaps an angel of the Lord surveyed an endless sea of chaos,
then troubled it gently with his finger. In this tiny and temporary swirl of
equations, our cosmos took shape."

—MARTIN GARDNER, "ORDER AND SURPRISE," *PHILOSOPHY OF SCIENCE*, 1950

FEBRUARY 13

"I'm thinking about a fourth spatial dimension, like length, breadth, and thickness.
For economy of materials and convenience of arrangement you couldn't beat it.
To say nothing of the saving of ground space—you could put an eight-room house
on the land now occupied by a one-room house."

—ROBERT HEINLEIN, "AND HE BUILT A CROOKED HOUSE," 1940

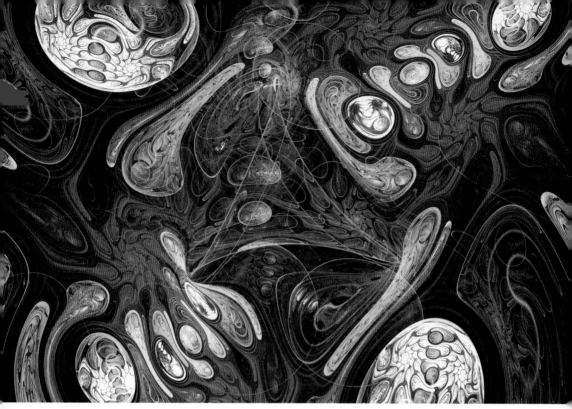

FEBRUARY 14

"Our mathematical models of physical reality are far from complete,
but they provide us with schemes that model reality with great precision—
a precision enormously exceeding that of any description that is free of mathematics."

—ROGER PENROSE, "WHAT IS REALITY?" *NEW SCIENTIST*, 2006

FEBRUARY 15

Galileo Galilei, 1564

"Lest we forget: Galileo, the greatest scientist of his time, was forced to his knees
under threat of torture and death, obliged to recant his understanding of the
Earth's motion, and placed under house arrest for the rest of his life
by steely-eyed religious maniacs. He worked at a time when every European
intellectual lived in the grip of a Church that thought nothing of burning scholars
alive for merely speculating about the nature of the stars. . . .
This is the same Church that did not absolve Galileo of heresy for 350 years (in 1992)."

—SAM HARRIS, "THE LANGUAGE OF IGNORANCE," *TRUTHDIG*, 2006

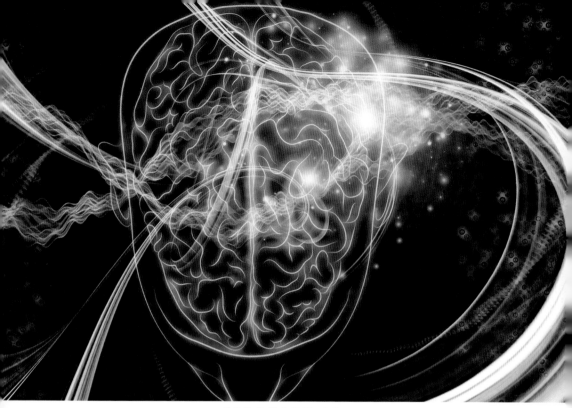

FEBRUARY 16

"There is no reason why the most fundamental aspects of the laws of nature should be within the grasp of human minds . . . nor why those laws should have testable consequences at the moderate energies and temperatures [of life-bearing planets]. . . . As we probe deeper into . . . the nature of reality, we can expect to find more of these deep results which limit what can be known. Ultimately, we may even find that their totality characterizes the universe more precisely than the catalogue of those things that we can know."

—JOHN BARROW, *BOUNDARIES AND BARRIERS*, 1996

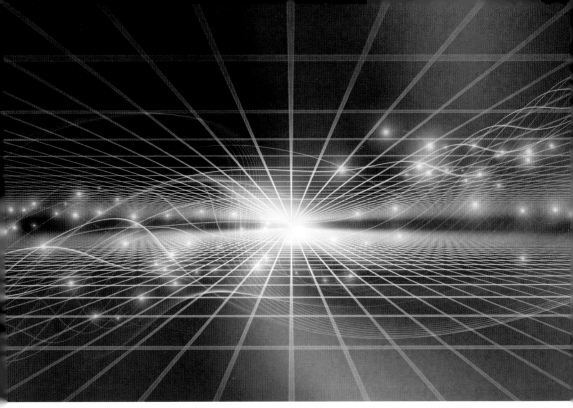

FEBRUARY 17

"I like the stars. It's the illusion of permanence, I think. I mean,
they're always flaring up and caving in and going out. But from here,
I can pretend. . . . I can pretend that things last. I can pretend that lives last longer
than moments. Gods come, and gods go. Mortals flicker and flash and fade.
Worlds don't last; and stars and galaxies are transient, fleeting things that twinkle
like fireflies and vanish into cold and dust. But I can pretend."

—NEIL GAIMAN, *THE SANDMAN* #48: "BRIEF LIVES 8, JOURNEY'S END," 1993

FEBRUARY 18

BORN ON THIS DAY: Alessandro Volta, 1745; Ernst Mach, 1838

"Isaac Newton discovered laws of motion that apply equally to
a planet moving through space and to an apple falling earthward,
revealing that the physics of the heavens and the earth are one.
Two hundred years later, Michael Faraday and James Clerk Maxwell
showed that electric currents produce magnetic fields,
and moving magnets can produce electric currents, establishing that
these two forces are as united as Midas' touch and gold."

—BRIAN GREENE, "THE UNIVERSE ON A STRING," *NEW YORK TIMES,* 2006

FEBRUARY 19

BORN ON THIS DAY: Nicolaus Copernicus, 1473

"Galileo championed a view of the universe, Copernicus's [Sun-centered universe],
that seemed not only new but shocking. Many churchmen who had never
even heard of Copernicus now learned that he had fathered these disturbing ideas.
An Italian bishop wanted Copernicus thrown in jail and was surprised
to learn that he had been dead for seventy years."

—JAMES C. DAVIS, *THE HUMAN STORY*, 2004

FEBRUARY 20

BORN ON THIS DAY: Ludwig Boltzmann, 1844

"Amazingly, this process of generating entropy is universal. It is what happens when a candle burns, when the sun shines, and when your stomach digests your lunch. In every instance, there is an inexorable, irreversible trend toward disorder and an increase in the total amount of information in the world."

—COREY S. POWELL, "WELCOME TO THE MACHINE," *NEW YORK TIMES*, 2006

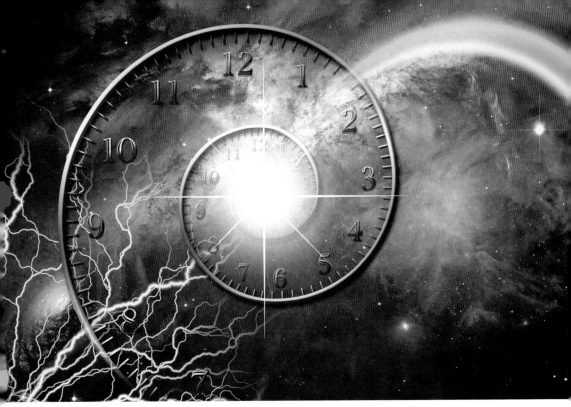

FEBRUARY 21

"With our eyes we can go back and forth over a time axis from
1790 to the present, but . . . we cannot pull such a stunt in human time.
In relativity theory, in the subtle fusion of time and space
known as Minkowskian space-time, the space dimensions
seem to lord it over the time dimensions, and the
whole structure exists as a frozen manifold outside of time."

—PHILIP J. DAVIS AND REUBEN HERSH, *DESCARTES' DREAM*, 1986

FEBRUARY 22

BORN ON THIS DAY: Heinrich Hertz, 1857

"Instruments enlarge the senses and make them more precise and reliable;
[physicist Robert] Hooke speaks of 'supplying of their infirmities with instruments,
and, as it were, the adding of artificial organs to the natural.'
He included here not only the obvious examples like microscopes and telescopes,
but also instruments related, say, to magnetism,
being used to investigate a phenomenon not directly sensible at all."

—JIM BENNETT, "ROBERT HOOKE AS MECHANIC AND NATURAL PHILOSOPHER,"
NOTES AND RECORDS OF THE ROYAL SOCIETY OF LONDON, 1980

FEBRUARY 23

"Gödel proved that the world of pure mathematics is inexhaustible; no finite set of axioms and rules of inference can ever encompass the whole of mathematics; given any finite set of axioms, we can find meaningful mathematical questions which the axioms leave unanswered. I hope that an analogous situation exists in the physical world. If my view of the future is correct, it means that the world of physics and astronomy is also inexhaustible; no matter how far we go into the future, there will always be new things happening, new information coming in, new worlds to explore, a constantly expanding domain of life, consciousness, and memory."

—FREEMAN J. DYSON, "TIME WITHOUT END: PHYSICS AND BIOLOGY IN AN OPEN UNIVERSE," *REVIEWS OF MODERN PHYSICS*, 1979

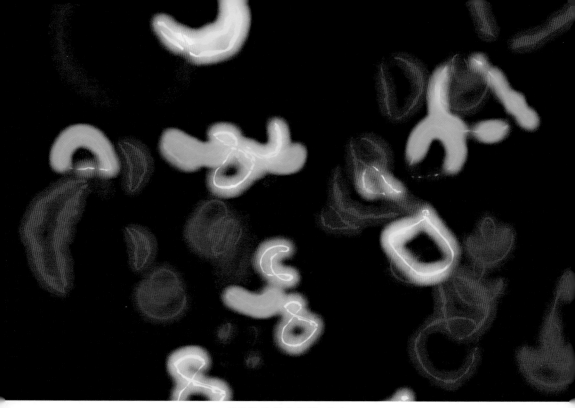

FEBRUARY 24

"Back in the early 1970s, one of the pioneers of string theory,
Italian physicist Daniele Amati, characterized the theory as
'part of 21st-century physics that fell by chance into the 20th century.'
He meant that string theory had been invented by chance and developed
by a process of tinkering, without physicists really grasping what was behind it."

—EDWARD WITTEN, "CAN SCIENTISTS' 'THEORY OF EVERYTHING'
REALLY EXPLAIN ALL THE WEIRDNESS THE UNIVERSE DISPLAYS?," *ASTRONOMY*, 2002

FEBRUARY 25

"The unbelievably small and the unbelievably vast eventually meet—like the closing
of a gigantic circle. I looked up, as if somehow I would grasp the heavens.
The universe, worlds beyond number, God's silver tapestry spread across the night.
And in that moment, I knew the answer to the riddle of the infinite. I had thought
in terms of man's own limited dimension. I had presumed upon nature. That existence
begins and ends in man's conception, not nature's. And I felt my body dwindling,
melting, becoming nothing. My fears melted away. And in their place
came acceptance. All this vast majesty of creation, it had to mean something.
And then I meant something, too. Yes, smaller than the smallest,
I meant something, too. To God, there is no zero. I still exist!"

—SCOTT CAREY, CHARACTER IN THE MOVIE *THE INCREDIBLE SHRINKING MAN*, 1957

FEBRUARY 26

"In 1676, Isaac Newton explained his accomplishments through a simple metaphor. 'If I have seen farther it is by standing on the shoulders of giants,' he wrote. The image wasn't original to him, but in using it Newton reinforced a way of thinking about scientific progress that remains popular: We learn about the world through the vision of a few colossal figures."

—PETER DIZIKES, "TWILIGHT OF THE IDOLS," *NEW YORK TIMES BOOK REVIEW*, 2006

FEBRUARY 27

"Scientists are remarkably sloppy about their use of the word 'law.' . . .
It would be nice, for example, if something that had been verified a thousand times
was called an 'effect,' something verified a million times a 'principle,' and something
verified 10 million times a 'law,' but things just don't work that way.
The use of these terms is based entirely on historical precedent and
has nothing to do with the confidence scientists place in a particular finding."

—JAMES TREFIL, *THE NATURE OF SCIENCE*, 2003

FEBRUARY 28

"The Christians know that the mathematical principles according to which the corporeal world was to be created are coeternal with God. . . . Geometry . . . has supplied God with the models for the creation of the world. With the image of God it has passed into man, and was certainly not received within through the eyes."

—JOHANNES KEPLER, *HARMONICES MUNDI (THE HARMONY OF THE WORLD)*, 1619

FEBRUARY 29

"It seemed that Earth was being taken and remade not by ETs from another spiral arm of the Milky Way or from another galaxy, but by beings from another *universe*, where all the laws of nature were radically different from those in this one. Humanity's reality, which operated on Einsteinian laws, and the utterly different reality of humanity's dispossessors had collided, meshed. At this Einstein intersection, all things seemed possible now in this worst of all possible new worlds."

—DEAN KOONTZ, *THE TAKING*, 2004

MARCH 1

"It is interesting to note the curious mental attitude of scientists working on 'hopeless' subjects. Contrary to what one might at first expect, they are all buoyed up by irrepressible optimism. I believe there is a simple explanation for this. Anyone without such optimism simply leaves the field and takes up some other line of work. Only the optimists remain."

—FRANCIS CRICK, *WHAT MAD PURSUIT*, 1988

MARCH 2

"Bright star, would I were steadfast as thou art
Not in lone splendour hung aloft the night
And watching, with eternal lids apart,
Like Nature's patient, sleepless Eremite,
The moving waters at their priestlike task
Of pure ablution round earth's human shores,
Or gazing on the new soft-fallen mask
Of snow upon the mountains and the moors."

—JOHN KEATS "BRIGHT STAR, WOULD I WERE STEADFAST," 1838

MARCH 3

"Recent observations and experiments suggest that our universe is simple.
The distribution of matter and energy is remarkably uniform.
The hierarchy of complex structures, ranging from galaxy clusters to subnuclear
particles, can be described in terms of a few dozen elementary constituents
and less than a handful of forces, all related by simple symmetries.
A simple universe demands a simple explanation."

—PAUL J. STEINHARDT, IN JOHN BROCKMAN'S *WHAT WE BELIEVE BUT CANNOT PROVE*, 2006

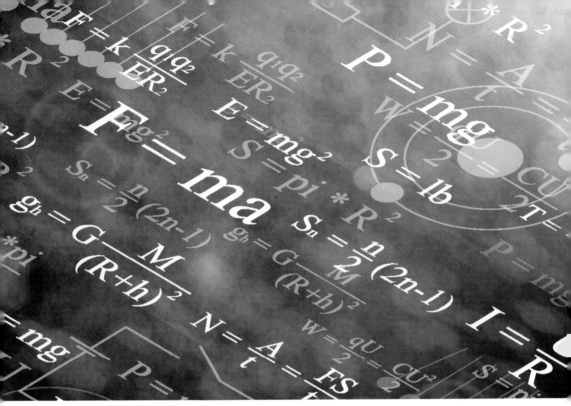

MARCH 4

"In 2056, I think you'll be able to buy a T-shirt on which
are printed equations describing the unified physical laws of our universe.
All the laws we have discovered so far will be
derivable from these equations."

—MAX TEGMARK, "MAX TEGMARK FORECASTS THE FUTURE," *NEW SCIENTIST*, 2006

MARCH 5

"Time was when all scientists were outsiders. Self-funded or backed by a rich benefactor, they pursued their often wild ideas in home-built labs with no one to answer to but themselves. From Nicolaus Copernicus to Charles Darwin, they were so successful that it's hard to imagine what modern science would be like without them. Their isolated, largely unaccountable ways now seem the antithesis of modern science, with consensus and peer review at its very heart."

— "IT PAYS TO KEEP A LITTLE CRAZINESS," EDITORIAL, *NEW SCIENTIST*, 2006

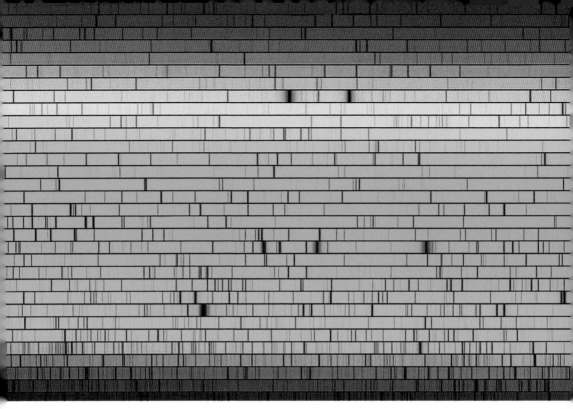

MARCH 6

BORN ON THIS DAY: Joseph Fraunhofer, 1787

"Spectrum analysis, which, as we hope we have shown, offers a wonderfully simple means for discovering the smallest traces of certain elements in terrestrial substances, also opens to chemical research a hitherto completely closed region extending far beyond the limits of the earth and even of the solar system. Since in this analytical method it is sufficient to see the glowing gas to be analyzed, it can easily be applied to the atmosphere of the sun and the bright stars."

—GUSTAV KIRCHHOFF AND ROBERT BUNSEN,
"CHEMICAL ANALYSIS BY OBSERVATION OF SPECTRA," *ANNALEN DER PHYSIK UND DER CHEMIE*
(*ANNALS OF PHYSICS AND CHEMISTRY*), 1860

MARCH 7

"First, [Newton's Law of Universal Gravitation] is mathematical in its expression. . . .
Second, it is not exact; Einstein had to modify it. . . .
There is always an edge of mystery, always a place where we have some fiddling
around to do yet. . . . But the most impressive fact is that gravity is simple. . . .
It is simple, and therefore it is beautiful. . . . Finally, comes the universality
of the gravitational law and the fact that it extends over such enormous distances. . . ."

—RICHARD FEYNMAN, *THE CHARACTER OF PHYSICAL LAW*, 1965

MARCH 8

"If one could travel in time, what wish could not be answered?
All the treasures of the past would fall to one man with a submachine gun.
Cleopatra and Helen of Troy might share his bed, if bribed
with a trunkful of modern cosmetics."

—LARRY NIVEN, "THE THEORY AND PRACTICE OF TIME TRAVEL," IN *ALL THE MYRIAD WAYS*, 1971

MARCH 9

"The Muslims were the leading scholars between the seventh and fifteenth centuries, and were the heirs of the scientific traditions of Greece, India, and Persia. . . . [Scientific] activities were cosmopolitan, in that the participants were Arabs, Persians, Central Asians, Christians, and Jews, and later included Indians and Turks. The transfer of the knowledge of Islamic science to the West . . . paved the way for the Renaissance, and for the scientific revolution in Europe."

—MAURIZIO IACCARINO, "SCIENCE AND CULTURE," *EMBO REPORTS*, 2003

MARCH 10

"Today there is a wide measure of agreement, which on the physical side of science approaches almost to unanimity, that the stream of knowledge is heading towards a non-mechanical reality; the universe begins to look more like a great thought than like a great machine. Mind no longer appears as an accidental intruder into the realm of matter; we are beginning to suspect that we ought rather to hail it as a creator and governor of the realm of matter. . . ."

—JAMES HOPWOOD JEANS, *THE MYSTERIOUS UNIVERSE*, 1930

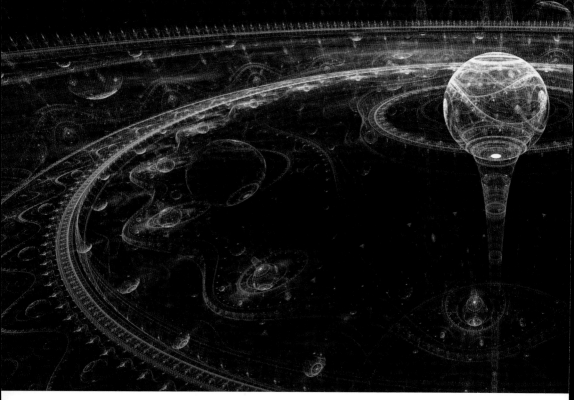

MARCH 11

"The universe we inhabit, and its operational principles, exist independently of our
observation or understanding; mathematical models of the universe . . .
are descriptive tools that exist only in our minds. Mathematics is at root
a formal description of orderliness, and since the universe is orderly
(at least on scales of space-time . . . which [we can] observe),
it should come as no surprise that the real world is well modeled mathematically."

—KEITH BACKMAN, "THE DANGER OF MATHEMATICAL MODELS," *SCIENCE*, 2006

MARCH 12

"It would be entirely wrong to suggest that science is something that knows everything already. Science proceeds by having hunches, by making guesses, by having hypotheses, sometimes inspired by poetic thoughts, by aesthetic thoughts even, and then science goes about trying to demonstrate it experimentally or observationally. And that's the beauty of science, that it has this imaginative stage but then it goes on to the proving stage, the demonstrating stage."

—RICHARD DAWKINS, IN JOHN BROCKMAN'S *WHAT WE BELIEVE BUT CANNOT PROVE*, 2006

MARCH 13

"[Newton's Laws of Motion] form the basis not only of classical dynamics, but of classical physics in general. Although they involve certain definitions and can in a sense be regarded as axioms, Newton asserted that they are based on quantitative observation and experiment; certainly they cannot be derived from other more basic relationships. The test of their validity involves predictions. . . . The validity of such predictions was verified in every case for more than two centuries."

—DUDLEY WILLIAMS AND JOHN SPANGLER, *PHYSICS FOR SCIENCE AND ENGINEERING*, 1981

MARCH 14

BORN ON THIS DAY: Albert Einstein, 1879

"Einstein's equations, in some sense, were like a Trojan horse. On the surface, the horse looks like a perfectly acceptable gift, giving us the observed bending of starlight under gravity and a compelling explanation of the origin of the universe. However, inside lurk all sorts of strange demons and goblins, which allow for the possibility of interstellar travel through wormholes and time travel. The price we had to pay for peering into the darkest secrets of the universe was the potential downfall of some of our most commonly held beliefs about our world— that its space is simply connected and its history is unalterable."

—MICHIO KAKU, *HYPERSPACE*, 1995

MARCH 15

"As measured by the millions of those who speak it fluently . . . ,
mathematics is arguably the most successful global language ever spoken. . . .
Equations are like poetry: They speak truths with a unique precision,
convey volumes of information in rather brief terms. . . . And just as conventional
poetry helps us to see deep *within* ourselves, mathematical poetry
helps us to see far *beyond* ourselves. . . ."

—MICHAEL GUILLEN, *FIVE EQUATIONS THAT CHANGED THE WORLD*, 1995

78

MARCH 16

"A century ago, the science and practice of electrical measurement . . .
hardly existed. With a few exceptions, ill-defined expressions relating to quantity and
intensity, combined with immature ideas of conductivity and derived circuits, retarded
the progress of quantitative electrical investigations. Yet, amidst this confusion,
a discovery had been made that was destined to convert order out of chaos,
to convert electrical measurement into the most precise of all physical operations,
and to aid almost every other branch of quantitative research. This discovery resulted
from the arduous labours of Georg Simon Ohm [1789–1854]."

—ROLLO APPLEYARD, *PIONEERS OF ELECTRICAL COMMUNICATION*, 1930

MARCH 17

"Making a real alteration in the time flow is a difficult thing.
You have to do something big, like killing a monarch. Simply being here,
I introduce tiny changes, but they are damped out by ten centuries of time,
and no real changes result down the line."

—ROBERT SILVERBERG, *UP THE LINE*, 1969

$$A(v) = \gamma(v)\begin{pmatrix} 1 & v \\ \alpha v & 1 \end{pmatrix}$$

$$\gamma(v) = \frac{1}{\sqrt{1-\alpha v^2}}$$

$$A'^0 = \frac{A^0 - \frac{v}{c}A^1}{\sqrt{1-\frac{v^2}{c^2}}}$$

$$E^2 - p^2 c = m^2 c^4$$

$$p = \frac{E}{c^2} u$$

$$\Delta t = \frac{u}{c^2} \Delta x$$

$$g_1 = \frac{(r\bar{b} + b\bar{r})}{\sqrt{2}}$$

$$g_4 = \frac{(r\bar{g} + g\bar{r})}{\sqrt{2}}$$

$$g_{\alpha\beta} = \begin{pmatrix} 1 & 0 & 0 & 0 \\ 0 & -1 & 0 & 0 \\ 0 & 0 & -1 & 0 \\ 0 & 0 & 0 & -1 \end{pmatrix}$$

$$\frac{u}{\sqrt{1 - v^2/c^2}}$$

$$\Delta s^2 = c^2 \Delta t^2 - \Delta x^2 - \Delta y^2 - \Delta z^2$$

$$v = v_0 \sqrt{\frac{c-v}{c+v}}$$

MARCH 18

"I do not agree with the view that the universe is a mystery. . . .
I feel that this view does not do justice to the scientific revolution that was started
almost four hundred years ago by Galileo and carried on by Newton. They showed
that at least some areas of the universe . . . are governed by precise mathematical laws.
Over the years since then, we have extended the work of Galileo and Newton. . . .
We now have mathematical laws that govern everything we normally experience."

—STEPHEN HAWKING, *BLACK HOLES AND BABY UNIVERSES AND OTHER ESSAYS*, 1993

MARCH 19

"The second law of thermodynamics is, without a doubt, one of the most perfect laws in physics. Any reproducible violation of it, however small, would bring the discoverer great riches as well as a trip to Stockholm. The world's energy problems would be solved at one stroke. . . . Not even Maxwell's laws of electricity or Newton's law of gravitation are so sacrosanct, for each has measurable corrections coming from quantum effects or general relativity. The law has caught the attention of poets and philosophers and has been called the greatest scientific achievement of the nineteenth century. Engels disliked it, for it supported opposition to Dialectical Materialism, while Pope Pius XII regarded it as proving the existence of a higher being."

—IVAN P. BAZAROV, *THERMODYNAMICS*, 1964

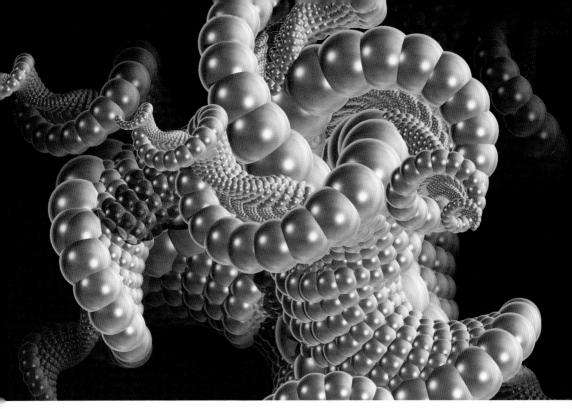

MARCH 20

"Mathematics, rightly viewed, possesses not only truth, but supreme beauty—
a beauty cold and austere, like that of sculpture."

—BERTRAND RUSSELL, *MYSTICISM AND LOGIC*, 1918

MARCH 21

BORN ON THIS DAY: Joseph Fourier, 1768

"Electrical resistance in cables and conductors can lead to burnt varnish, smoke, sudden short circuits, and melted metal; but without the benefit of the damping provided by resistance, without even the vestige of Joule heating . . . our machines might be super-efficient, but they would be afflicted with the mechanical equivalent of Parkinson's disease. . . . Without resistance, our electric blankets, kettles, and incandescent lamp bulbs would be useless."

—ANTONY ANDERSON, "SPARE A THOUGHT FOR THE OHM," *NEW SCIENTIST*, 1987

MARCH 22

BORN ON THIS DAY: Robert Millikan, 1868

"Long before Christians had come to believe in the Father, Son, and Holy Ghost, natural philosophers had stumbled on their own trinity: electricity, magnetism, and gravitational force. These three forces alone had governed the creation of the universe, they believed, and would shape its future forevermore. . . . Given the forces' disparate behaviors, it was no wonder that philosophers very early on were left scratching their heads: Were these three forces completely different? Or were they, like the Christian Trinity, three aspects of a single phenomenon?"

—MICHAEL GUILLEN, *FIVE EQUATIONS THAT CHANGED THE WORLD*, 1995

MARCH 23

"We just have a few fragments of Heraclitus—a hundred or so,
left behind like the bones of some fabulous beast . . .
[He wrote,] 'Eternity is a child playing checkers.'"

–STEPHEN MITCHELL, *THE ENLIGHTENED MIND*, 1991

MARCH 24

"[Some claim that] we do not understand why the universe has rules;
therefore, God must have done it. . . . [This argument] ignores the possibility
that the universe has to have rules, or could not exist. Or the fact that,
had it no rules, we would not be able to exist."

—BEN HOSKIN, "GOD OF THE GAPS," LETTER TO *NEW SCIENTIST*, 2007

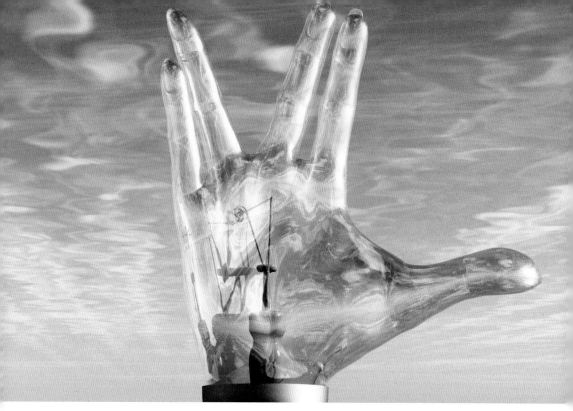

MARCH 25

"I canna change the laws of physics, Captain!"

—"SCOTTY" MONTGOMERY SCOTT TO CAPTAIN KIRK, "THE NAKED TIME," *STAR TREK* TV SERIES, 1966

MARCH 26

"I believe that the clue to [Isaac Newton's] mind is to be found in his unusual powers of continuous concentrated introspection. . . . His peculiar gift was the power of holding continuously in his mind a purely mental problem until he had seen straight through it. I fancy his pre-eminence is due to his muscles of intuition being the strongest and most enduring with which a man has ever been gifted. . . . I believe that Newton could hold a problem in his mind for hours and days and weeks until it surrendered to him its secret."

—JOHN MAYNARD KEYNES, "ESSAYS IN BIOGRAPHY: NEWTON, THE MAN,"
THE COLLECTED WRITINGS OF JOHN MAYNARD KEYNES, 1972

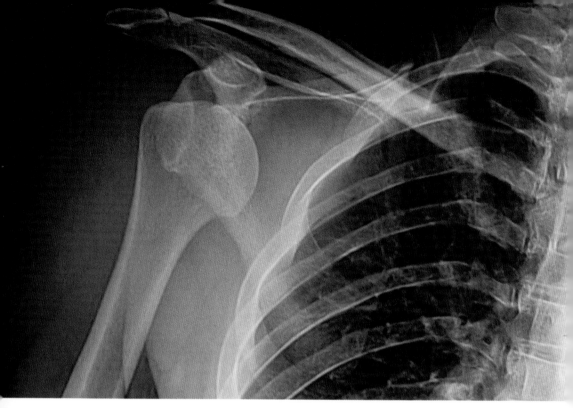

MARCH 27

BORN ON THIS DAY: Wilhelm Röntgen, 1845

"As the nineteenth century drew to a close, scientists could reflect with satisfaction that they had pinned down most of the mysteries of the physical world: electricity, magnetism, gases, optics, acoustics, kinetics, and statistical mechanics . . . all had fallen into order before them. They had discovered the X ray, the cathode ray, the electron, and radioactivity, invented the ohm, the watt, the Kelvin, the joule, the amp, and the little erg."

—BILL BRYSON, *A SHORT HISTORY OF NEARLY EVERYTHING*, 2004

MARCH 28

"It is indeed a surprising and fortunate fact that nature can be expressed by relatively low-order mathematical functions."

—RUDOLF CARNAP, CLASSROOM LECTURE,
QUOTED IN MARTIN GARDNER'S "ORDER AND SURPRISE," *PHILOSOPHY OF SCIENCE*, 1950

MARCH 29

"To a frog with its simple eye, the world is a dim array of grays and blacks.
Are we like frogs in our limited sensorium, apprehending just part
of the universe we inhabit? Are we as a species now awakening to the reality of
multidimensional worlds in which matter undergoes subtle reorganizations
in some sort of hyperspace?"

—MICHAEL MURPHY, *THE FUTURE OF THE BODY*, 1992

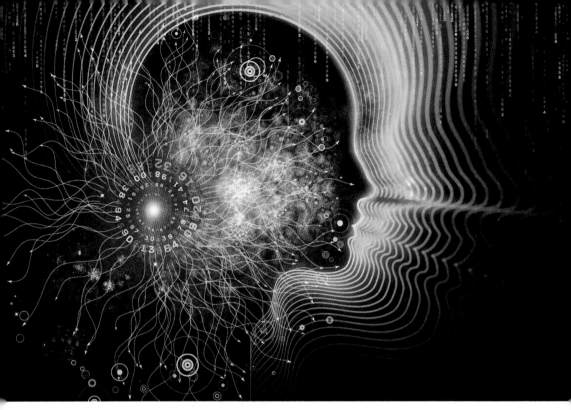

MARCH 30

"As the island of knowledge grows, the surface that makes contact with mystery expands. When major theories are overturned, what we thought was certain knowledge gives way, and knowledge touches upon mystery differently. This newly uncovered mystery may be humbling and unsettling, but it is the cost of truth. Creative scientists, philosophers, and poets thrive at this shoreline."

—W. MARK RICHARDSON, "A SKEPTIC'S SENSE OF WONDER," *SCIENCE*, 1998

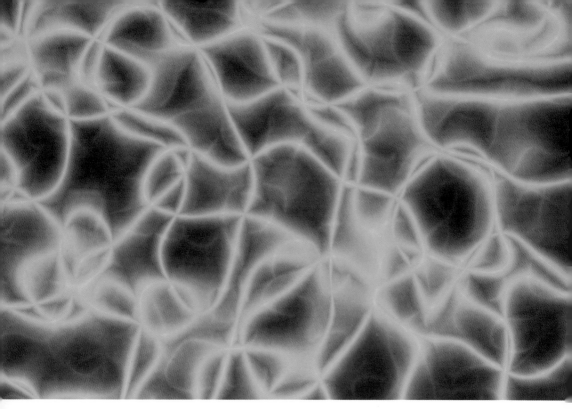

MARCH 31

"If reality really is vested in the present, then you have the power to change that reality across the universe, back and forth in time, by simple perambulation. But, then, so does an Andromedan sentient green blob. If the blob oozes to the left and then the right, the present moment on Earth (as judged by the blob, in its frame of reference) will lurch though huge changes back and forth in time."

—PAUL DAVIES, *ABOUT TIME*, 1995

APRIL 1

"It often happens in science that two researchers will hit upon the same solution
to a problem simultaneously and independently, usually because
both are working on the same problem and following along the same path.
As Mark Twain put it, 'When it's steamboat time, you steam.'"

—BEN BOVA, *THE STORY OF LIGHT*, 2001

APRIL 2

"The system which Galileo advocated was the orthodox Copernican system, designed . . . nearly a century before Kepler threw out the epicycles. . . . Incapable of acknowledging that any of his contemporaries had a share in the progress of astronomy, Galileo blindly and indeed suicidally ignored Kepler's work to the end, persisting in the futile attempt to bludgeon the world into accepting a Ferris wheel with forty-eight epicycles as 'rigorously demonstrated' physical reality."

—ARTHUR KOESTLER, *THE SLEEPWALKERS: A HISTORY OF MAN'S CHANGING VISION OF THE UNIVERSE*, 1959

APRIL 3

"It is a mark of Newton's genius that of all the possible statements about motion,
he recognized that three and only three completely define a logically
consistent framework within which all problems of motion
can be analyzed quantitatively. These are Newton's three laws."

—ERNEST S. ABERS AND CHARLES F. KENNEL, *MATTER IN MOTION*, 1977

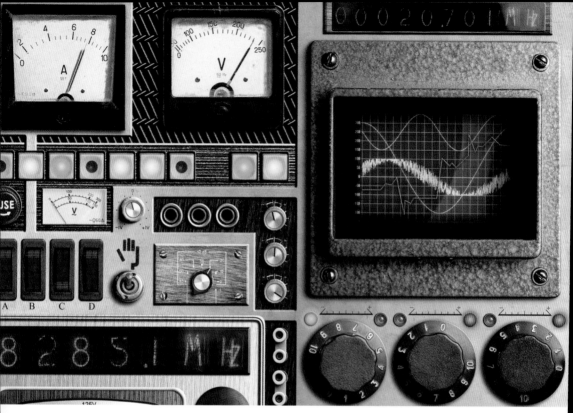

APRIL 4

"Scientific principles and laws do not lie on the surface of nature.
They are hidden, and must be wrested from nature by
an active and elaborate technique of inquiry."

—JOHN DEWEY, *RECONSTRUCTION IN PHILOSOPHY*, 1920

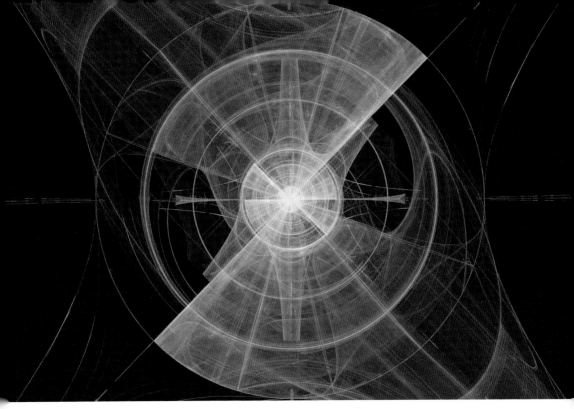

APRIL 5

"Hyperbolic geometry, conceived by mathematician Carl Gauss in 1816, . . . describes a world that is curving away from itself at every point, making it the precise opposite of a sphere. . . . Gauss never published the idea, perhaps because he found it inelegant. In 1825 the Hungarian mathematician János Bolyai and the Russian mathematician Nikolay Lobachevsky independently rediscovered hyperbolic geometry."

—DAVID SAMUELS, "KNIT THEORY," *DISCOVER*, 2006

APRIL 6

"A large number of areas of the brain are involved when viewing equations, but when one looks at a formula rated as beautiful it activates the emotional brain— the medial orbito-frontal cortex—like looking at a great painting or listening to a piece of music. Neuroscience can't tell you what beauty is, but if you find it beautiful the medial orbito-frontal cortex is likely to be involved; you can find beauty in anything."

—SEMIR ZEKI, QUOTED IN JAMES GALLAGHER'S,
"MATHEMATICS: WHY THE BRAIN SEES MATHS AS BEAUTY," *BBC NEWS*, 2014

APRIL 7

"For the religious, passivism [i.e., the view that objects are obedient to the laws of nature] provides a clear role for God as the author of the laws of nature. If the laws of nature are God's commands for an essentially passive world, . . . God also has the power to suspend the laws of nature, and so perform miracles."

—BRIAN DAVID ELLIS, *THE PHILOSOPHY OF NATURE: A GUIDE TO THE NEW ESSENTIALISM*, 2002

APRIL 8

"This solution to the problem of induction involves accepting the existence of laws of nature, and it involves recognizing these laws not just as regularities in the behavior of things (consistencies in how the world works in different places and times), but as forms of natural necessity—as laws whose obtaining ensures that things behave and interact in certain regular ways."

—JOHN FOSTER, *THE DIVINE LAWMAKER: LECTURES ON INDUCTION, LAWS OF NATURE, AND THE EXISTENCE OF GOD*, 2004

APRIL 9

"The fundamental advances in aerodynamics in the eighteenth century began
with the work of Daniel Bernoulli (1700–1782). Newtonian mechanics
had unlocked, but not opened, the door to modern hydrodynamics.
Bernoulli was the first to open that door, though just by a crack.
Leonhard Euler and others who would follow would fling the door wide open."

—JOHN D. ANDERSON JR., *A HISTORY OF AERODYNAMICS: AND ITS IMPACT ON FLYING MACHINES*, 1997

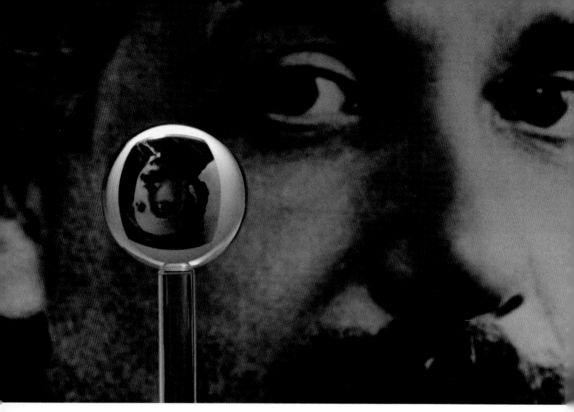

APRIL 10

"Until Einstein's time, scientists typically would observe things, record them, then find a piece of mathematics that explained the results. [Sylvester James Gates says,] 'Einstein exactly reverses that process. He starts off with a beautiful piece of mathematics that's based on some very deep insights into the way the universe works and then, from that, makes predictions about what ought to happen in the world. It's a stunning reversal to the usual ordering in which science is done. . . . [Einstein demonstrated] the power of human creativity in the sciences.'"

—PETER TYSON'S "THE LEGACY OF $E = MC^2$," *NOVA*, WWW.PBS.ORG, 2005

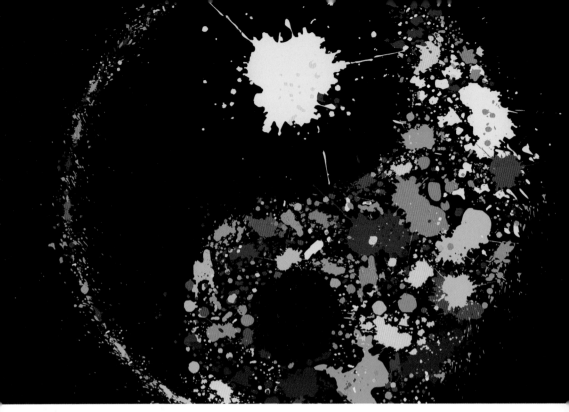

APRIL 11

"Science, like life, feeds on its own decay.
New facts burst old rules; then newly divined conceptions
bind old and new together into a reconciling law."

—WILLIAM JAMES, *THE WILL TO BELIEVE AND OTHER ESSAYS IN POPULAR PHILOSOPHY*, 1897

APRIL 12

"I have often made the hypothesis that ultimately
physics will not require a mathematical statement, that in the end
the machinery will be revealed, and the laws will turn out to be simple,
like the chequer board with all its apparent complexities."

—RICHARD FEYNMAN, *THE CHARACTER OF PHYSICAL LAW*, 1965

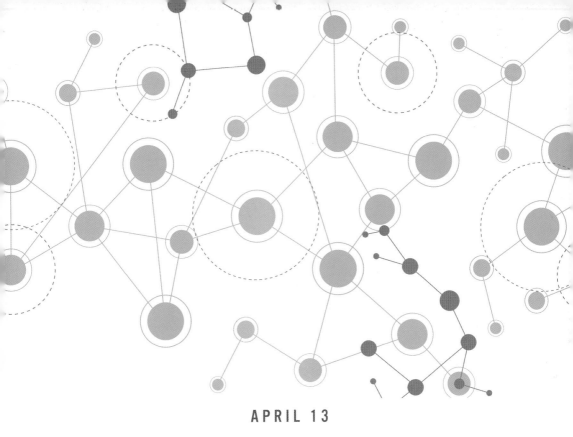

APRIL 13

"It is these connections that form the fabric of physics. It is the joy of the theoretical physicist to discover them, and of the experimentalist to test their strength. . . . In the end, what science does is change the way we think about the world and our place within it. . . . There is universal joy in making new connections."

—LAWRENCE M. KRAUSS, *FEAR OF PHYSICS*, 1993

APRIL 14

BORN ON THIS DAY: Christiaan Huygens, 1629

"I saw Eternity the other night,
Like a great ring of pure and endless light,
All calm, as it was bright;
And round beneath it, Time in hours, days, years,
Driv'n by the spheres
Like a vast shadow mov'd; in which the world
And all her train were hurl'd."

—HENRY VAUGHAN, "THE WORLD," 1650

APRIL 15

"The great equations of modern physics are a permanent part of scientific knowledge, which may outlast even the beautiful cathedrals of earlier ages."

—STEVEN WEINBERG, IN GRAHAM FARMELO'S *IT MUST BE BEAUTIFUL*, 2003

APRIL 16

"It is the most persistent and greatest adventure in human history, this search
to understand the universe, how it works and where it came from. It is difficult to
imagine that a handful of residents of a small planet circling an insignificant star in a
small galaxy have as their aim a complete understanding of the entire universe,
a small speck of creation truly believing it is capable of comprehending the whole."

—MURRAY GELL-MANN, IN JOHN BOSLOUGH'S *STEPHEN HAWKING'S UNIVERSE*, 1989

APRIL 17

"It should no longer seem strange that man, the ape of his Creator, has finally discovered how to sing polyphonically, an art unknown to the ancients. With this symphony of voices, man can play through the eternity of time in less than an hour and can taste in small measure the delight of God the Supreme Artist by calling forth that very sweet pleasure of the music that imitates God."

—JOHANNES KEPLER, *HARMONICES MUNDI (THE HARMONY OF THE WORLD)*, 1619

APRIL 18

"No other part of science has contributed as much to the liberation of the human spirit as the Second Law of Thermodynamics. Yet, at the same time, few other parts of science are held to be so recondite. Mention of the Second Law raises visions of lumbering steam engines, intricate mathematics, and infinitely incomprehensible entropy. Not many would pass C. P. Snow's test of general literacy, in which not knowing the Second Law is equivalent to not having read a work of Shakespeare."

—PETER W. ATKINS, *THE SECOND LAW*, 1984

APRIL 19

"Newton was the greatest creative genius physics has ever seen.
None of the other candidates for the superlative (Einstein, Maxwell, Boltzmann,
Gibbs, and Feynman) has matched Newton's combined achievements as theoretician,
experimentalist, and mathematician. . . . If you were to become a time traveler
and meet Newton on a trip back to the seventeenth century,
you might find him something like the performer who first exasperates
everyone in sight and then goes on stage and sings like an angel."

—WILLIAM H. CROPPER, *GREAT PHYSICISTS*, 2004

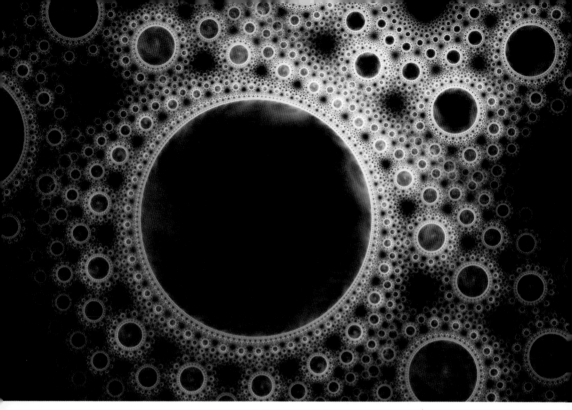

APRIL 20

"Natural selection gave us our ornately structured biosphere, and perhaps a similar evolutionary principle operates in the genesis of universes. Our universe may have arisen from selection for intelligences that can make fresh universes. . . . Selection arises because only firm laws can yield constant, benign conditions to form new life. Once life forms realize this, they could intentionally make more smart universes with the right fixed laws to produce ever grander structures."

—GREGORY BENFORD IN JOHN BROCKMAN'S *WHAT WE BELIEVE BUT CANNOT PROVE*, 2006

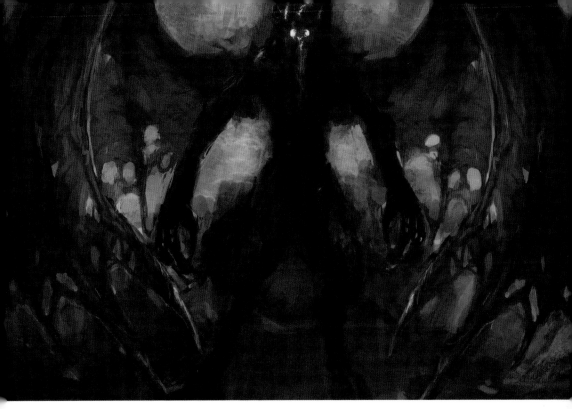

APRIL 21

"Enlightenment 'natural theology,' . . . presuming the Creator
to have had our best interests at heart when he instituted nature's laws
and then retired, made no allowance for either Satanic influence or
divine payback for wickedness. God's indifference . . . was more complete
than any deist had dared to conceive."

—FREDERICK C. CREWS, *FOLLIES OF THE WISE*, 2006

APRIL 22

"The empirical basis of objective science has nothing 'absolute' about it.
Science does not rest upon solid bedrock. The bold structure of its theories rises,
as it were, above a swamp. It is like a building erected on piles. The piles are driven
down from above into the swamp, but not down to any natural or 'given' base;
and if we stop driving the piles deeper, it is not because we have reached firm ground.
We simply stop when we are satisfied that the piles are firm enough to carry
the structure, at least for the time being."

—KARL POPPER, *THE LOGIC OF SCIENTIFIC DISCOVERY*, 1959

APRIL 23

BORN ON THIS DAY: Max Planck, 1858

"When in the year 5000 people look back three thousand years to our era, we all hope that they will find some epochal events as myth-making for them as the Trojan War for us. Many events of such lasting significance are to be found among the achievements of our twentieth century scientific revolution in physics. The very first of these revolutionary events . . . is Planck's invention of the energy quantum in 1900. . . . Planck opened the door to an utterly new, totally unanticipated, wonderfully strange and mysterious but absolutely necessary ultimate reality of the world. . . . We cannot know where it will lead and we cannot believe it will end."

—IAN DUCK AND E.C.G. SUDARSHAN, *100 YEARS OF PLANCK'S QUANTUM*, 2000

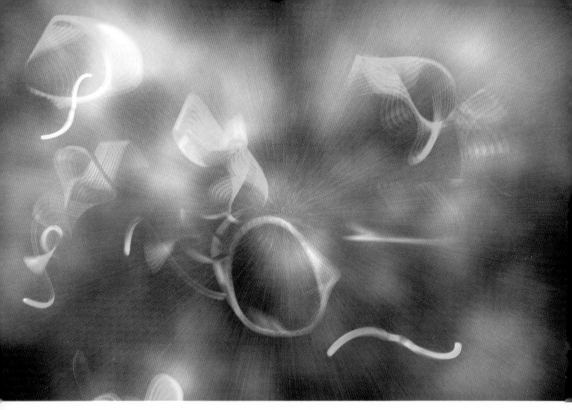

APRIL 24

"Professionally, I work on something called Superstring theory,
or now called M-theory, and the goal is to find an equation,
perhaps no more than one inch long, which will allow us to
'read the mind of God,' as Einstein used to say."

—MICHIO KAKU, "PARALLEL UNIVERSES, THE MATRIX, AND SUPERINTELLIGENCE," KURZWEILAI.NET, 2003

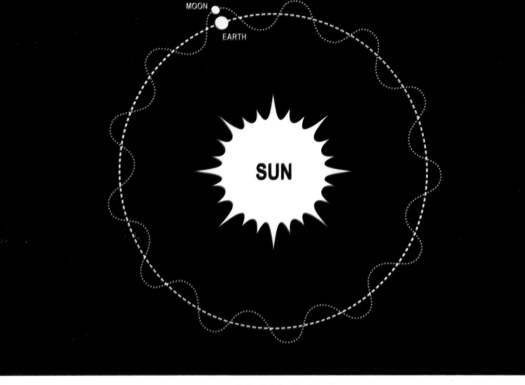

APRIL 25

BORN ON THIS DAY: Wolfgang Pauli, 1900

"The solutions [Heisenberg's matrix mechanics] provided came only with agony and labor and it demanded difficult concessions—for example, giving up the idea of 'orbits' within the atom. This aroused the wasp in Pauli: The moon, like an electron, occupied a stationary state, and yet it moved in an orbit. If nature made a place for orbits among the spheres, why did Heisenberg ban them from the atom and insist only on 'observables'? 'Physics is decidedly confused at the moment,' Pauli remarked in 1925. 'In any event, it is much too difficult for me and I wish I . . . had never heard of it.'"

—THOMAS POWERS, *HEISENBERG'S WAR: THE SECRET HISTORY OF THE GERMAN BOMB*, 2000

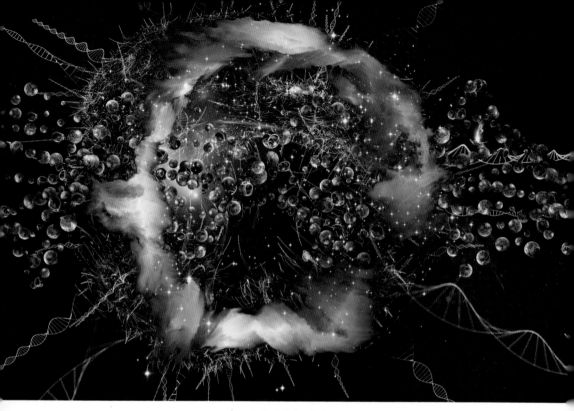

APRIL 26

"The laws of nature are the skeleton of the universe. They support it,
give it shape, tie it together. . . . They tell us that the universe is a place we can know,
understand, approach with the power of human reason. In an age that seems to be
losing confidence in its ability to manage things, they remind us that
even the most complex systems around us operate according to simple laws,
laws easily accessible to the average person."

—JAMES S. TREFIL, *THE NATURE OF SCIENCE*, 2003

APRIL 27

"That one body may act upon another at a distance through a vacuum without the mediation of anything else, by and through which their action and force may be conveyed from one to another, is to me so great an absurdity that, I believe, no man who has in philosophical matters a competent faculty of thinking can ever fall into it."

—ISAAC NEWTON, LETTER TO RICHARD BENTLEY, 1693

APRIL 28

"Science will continue to surprise us with what it discovers and creates;
then it will astound us by devising new methods to surprise us. At the core of science's
self-modification is technology. New tools enable new structures of knowledge and
new ways of discovery. The achievement of science is to know new things;
the evolution of science is to know them in new ways. What evolves is less the body
of what we know and more the nature of our knowing."

—KEVIN KELLY, "SPECULATIONS ON THE FUTURE OF SCIENCE," EDGE.ORG, 2006

APRIL 29

"A 'law of nature' is one of those concepts that slips through your fingers
the more you try to grasp it. The most that can be said about a physical law
is that it is a hypothesis that has been confirmed by experiment so many times
that it becomes universally accepted. There is nothing natural about it,
however: it is a wholly human construct."

—"BREAKING THE 'LAWS' OF NATURE," *NEW SCIENTIST*, EDITORIAL, 2006

APRIL 30

BORN ON THIS DAY: Carl Friedrich Gauss, 1777

"Mathematics is the gate and key of the sciences . . . Neglect of mathematics works injury to all knowledge, since he who is ignorant of it cannot know the other sciences or the things of this world. And what is worse, men who are thus ignorant are unable to perceive their own ignorance and so do not seek a remedy."

—ROGER BACON, *OPUS MAJUS*, 1266

MAY 1

"Truly the gods have not from the beginning revealed all things to mortals,
but by long seeking, mortals make progress in discovery."

—XENOPHANES OF COLOPHON (A SURVIVING FRAGMENT OF HIS TEXTS, C. 500 BCE)

MAY 2

"The trend of mathematics and physics towards unification provides the physicist with
a powerful new method of research into the foundations of his subject. . . .
The method is to begin by choosing that branch of mathematics which one thinks will
form the basis of the new theory. One should be influenced very much in this
choice by considerations of mathematical beauty. It would probably be a good thing
also to give a preference to those branches of mathematics that have an interesting
group of transformations underlying them, since transformations play an important role
in modern physical theory, both relativity and quantum theory seeming to show
that transformations are of more fundamental importance than equations."

—PAUL DIRAC, "THE RELATION BETWEEN MATHEMATICS AND PHYSICS,"
PROCEEDINGS OF THE ROYAL SOCIETY (EDINBURGH), 1938–1939

MAY 3

"Matter, though divisible in an extreme degree, is nevertheless not infinitely divisible. That is, there must be some point beyond which we cannot go in the division of matter . . . I have chosen the word 'atom' to signify these ultimate particles . . ."

—JOHN DALTON, *A NEW SYSTEM OF CHEMICAL PHILOSOPHY*, 1808

MAY 4

"As soon as we adventure on the paths of the physicist, we learn to weigh and to measure, to deal with time and space and mass and their related concepts, and to find more and more our knowledge expressed and our needs satisfied through the concept of number, as in the dreams of Plato and Pythagoras. . . ."

—D'ARCY WENTWORTH THOMPSON, *ON GROWTH AND FORM*, 1917

MAY 5

"You have only to stretch out your hand, close it quickly and you feel that you have caught mathematical air and that a few formulae are stuck to your palm. . . . Even the sun rays must remember, when passing through the windows, the laws to which they are subject according to the will of God, Newton, Einstein, and Heisenberg."

—LEOPOLD INFELD, *QUEST*, 2006

MAY 6

"It is evident that an acquaintance with natural laws means
no less than an acquaintance with the mind of God therein expressed."

—JAMES JOULE, "ADDRESS TO THE BRITISH ASSOCIATION
FOR THE ADVANCEMENT OF SCIENCE," 1873

MAY 7

"Our attempts at modeling physical reality normally consist of two parts:
(1) A set of local laws that are obeyed by the various physical quantities.
These are usually formulated in terms of differential equations. (2) Sets of boundary
conditions that tell us the state of some regions of the universe at a certain time. . . .
Many people would claim that the role of science is confined to
the first of these and that theoretical physics will have achieved its goal
when we have obtained a complete set of local physical laws."

—STEPHEN HAWKING, *BLACK HOLES AND BABY UNIVERSES AND OTHER ESSAYS*, 1993

MAY 8

"When a distinguished but elderly scientist states that something is possible,
he is almost certainly right. When he states that something is impossible,
he is very probably wrong."

—ARTHUR C. CLARKE, *PROFILES OF THE FUTURE*, 1962

MAY 9

"When one 'sees' a mathematical truth, one's consciousness breaks through
into this world of ideas. . . . One may take the view that
in such cases the mathematicians have stumbled upon 'works of God'."

—ROGER PENROSE, *THE EMPEROR'S NEW MIND*, 1989

MAY 10

"Why do the laws that govern [the universe] seem constant in time and always at work?
. . . One can imagine a universe in which laws are not truly lawful. Talk of miracles
does just this, invoking God to make things work. Physics aims to find the laws
instead, and hopes that they will be uniquely constrained, as when Einstein wondered
whether God had any choice when He made the universe."

—GREGORY BENFORD, IN JOHN BROCKMAN'S *WHAT WE BELIEVE BUT CANNOT PROVE*, 2006

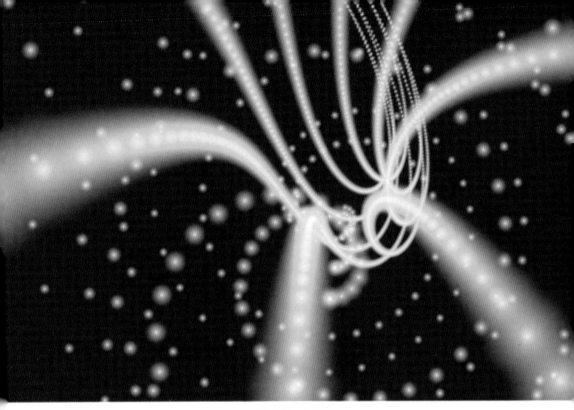

MAY 11

BORN ON THIS DAY: Richard Feynman, 1918

"I think I can safely say that nobody understands quantum mechanics."

—RICHARD FEYNMAN, *THE CHARACTER OF PHYSICAL LAW*, 1965

MAY 12

"Forty years as an astronomer have not quelled my enthusiasm for lying outside
after dark, staring up at the stars. It isn't only the beauty of the night sky
that thrills me. It's the sense I have that some of those points of light
are the home stars of beings not so different from us, daily cares and all,
who look across space with wonder, just as we do."

—FRANK DRAKE, *IS ANYONE OUT THERE?*, 1992

MAY 13

"He showed me a little thing, the quantity of a hazelnut,
in the palm of my hand, and it was as round as a ball. I looked thereupon
with the eye of my understanding and thought: What may this be?
And it was answered generally thus: It is all that is made."

—JULIAN OF NORWICH, *REVELATIONS OF DIVINE LOVE*, 1395

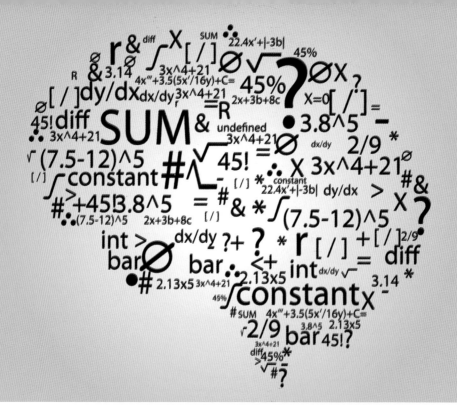

MAY 14

"Physicists had to invent words and phrases for concepts far removed from everyday experience. It was their fashion to avoid pure neologisms and instead to evoke, even if feebly, some analogous commonplace. The alternative was to name discoveries and equations after one another. This they did also. But if you didn't know it was physics they were talking, you might very well worry about them."

—CARL SAGAN, *CONTACT*, 1985

MAY 15

"Newton's law of universal gravitation was based on a variety of observations:
the paths of planets in their motion about the sun, the acceleration of objects near the
earth. . . . Physical laws are usually expressed as mathematical equations . . .
[that] can then be used to make predictions . . . It is usually easiest to learn the
physics and the necessary mathematics at about the same time since
the immediate application of mathematics to a physical situation helps
you understand both the physics and the mathematics."

—PAUL TIPLER, *PHYSICS*, 1976

MAY 16

"There is a noble vision of the great Castle of Mathematics, towering somewhere in the Platonic World of Ideas, which we humbly and devotedly discover (rather than invent). The greatest mathematicians manage to grasp outlines of the Grand Design, but even those to whom only a pattern on a small kitchen tile is revealed, can be blissfully happy. . . . Mathematics is a proto–text whose existence is only postulated but which nevertheless underlies all corrupted and fragmentary copies we are bound to deal with. The identity of the writer of this proto-text (or of the builder of the Castle) is anybody's guess. . . ."

—YURI I. MANIN, "MATHEMATICAL KNOWLEDGE: INTERNAL, SOCIAL, AND CULTURAL ASPECTS," *MATHEMATICS AS METAPHOR: SELECTED ESSAYS*, 2007

MAY 17

"The earth, that is sufficient,
I do not want the constellations any nearer,
I know they are very well where they are,
I know they suffice for those who belong to them."

—WALT WHITMAN, "SONG OF THE OPEN ROAD," *LEAVES OF GRASS*, 1856

MAY 18

"Taken to its ultimate, the cumulative audience paradox yields us
the picture of an audience of billions of time-travelers piled up in the past to witness
the Crucifixion, filling all the Holy Land and spreading out into Turkey, into Arabia,
even to India and Iran. . . . Yet at the original occurrence of those events,
no such hordes were present. . . . A time is coming, I thought, when we from down
the line will throng the past to the choking point. We will fill all our yesterdays
with ourselves and crowd out our own ancestors."

—ROBERT SILVERBERG, *UP THE LINE*, 1969

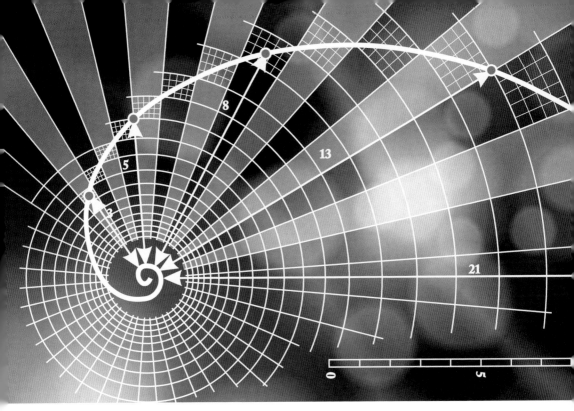

MAY 19

"The more important fundamental laws and facts of physical science have all been discovered, and these are now so firmly established that the possibility of their ever being supplanted in consequence of new discoveries is exceedingly remote."

—ALBERT MICHELSON, ADDRESS AT THE DEDICATION OF THE RYERSON PHYSICAL LABORATORY AT THE UNIVERSITY OF CHICAGO, 1894

MAY 20

"Where should we look to discover the principles that underpin reality?
In Einstein's view, while facts reside in the world, principles reside in the mind. . . .
Einstein insisted that great theories are those that explain the most facts
from the least number of principles. . . . The simpler the theory, the less it will look
anything like the world we see. The ultimate goal of science, he believed,
is to find one all-encompassing, self-evident principle or set of principles
from which the whole of reality can be deduced."

—AMANDA GEFTER, "HOW EINSTEIN PROBED THE POWER OF THE MIND," *NEW SCIENTIST*, 2005

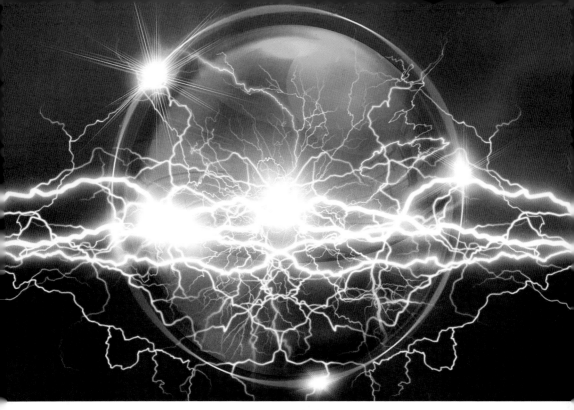

MAY 21

"I believe in science. Unlike mathematical theorems, scientific results can't be proved. They can only be tested again and again until only a fool would refuse to believe them. I cannot prove that electrons exist, but I believe fervently in their existence. And if you don't believe in them, I have a high-voltage cattle prod I'm willing to apply as an argument on their behalf. Electrons speak for themselves."

—SETH LLOYD, IN JOHN BROCKMAN'S *WHAT WE BELIEVE BUT CANNOT PROVE*, 2006

MAY 22

"When you discover mathematical structures that you believe correspond
to the world around you . . . you are communicating with the universe,
seeing beautiful and deep structures and patterns that no one without your training
can see. . . . The mathematics is there, it's leading you, and you are discovering it.
Mathematics is a profound language, an awesomely beautiful language.
For some, like Leibniz, it is the language of God. I'm not religious,
but I do believe that the universe is organized mathematically."

—ANTHONY TROMBA, QUOTED IN TIM STEPHENS'S "UCSC PROFESSOR SEEKS TO RECONNECT MATHEMATICS TO ITS
INTELLECTUAL ROOTS," UNIVERSITY OF CALIFORNIA PRESS RELEASE, 2003

MAY 23

John Bardeen, 1908

"Give me a place to stand on, and I will move the earth."

—ARCHIMEDES, UPON DISCOVERING THE PRINCIPLES OF LEVERS,
AS TOLD BY PAPPUS OF ALEXANDRIA, *SYNAGOGE (COLLECTION)*, BOOK VIII, C. 340

MAY 24

"Some time between the 13th and 15th centuries, Europe pulled well ahead
of the rest of the world in science and technology, a lead consolidated in the following
200 years. Then in 1687, Isaac Newton—foreshadowed by Copernicus, Kepler,
and others—had his glorious insight that the universe is governed by a few physical,
mechanical, and mathematical laws. This instilled tremendous confidence
that everything made sense, everything fitted together, and everything
could be improved by science."

—RICHARD KOCH AND CHRIS SMITH, "THE FALL OF REASON IN THE WEST," *NEW SCIENTIST*, 2006

MAY 25

"Pure mathematics and physics are becoming ever more closely connected, though their methods remain different. One may describe the situation by saying that the mathematician plays a game in which he himself invents the rules while the physicist plays a game in which the rules are provided by Nature, but as time goes on it becomes increasingly evident that the rules which the mathematician finds interesting are the same as those which Nature has chosen. . . . Possibly, the two subjects will ultimately unify, every branch of pure mathematics then having its physical application, its importance in physics being proportional to its interest in mathematics."

—PAUL DIRAC, "THE RELATION BETWEEN MATHEMATICS AND PHYSICS,"
PROCEEDINGS OF THE ROYAL SOCIETY (EDINBURGH), 1938–1939

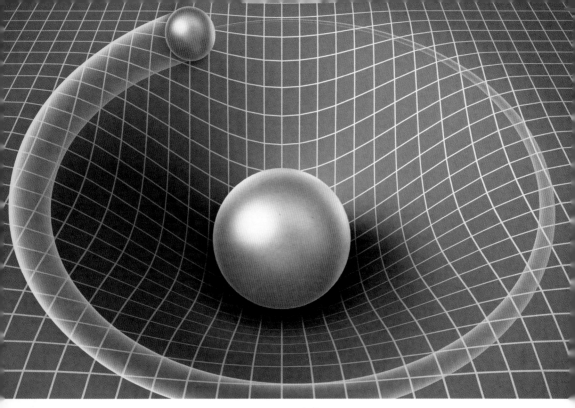

MAY 26

"It is gravity that binds us to the earth and that binds the earth and the other planets to the solar system. The gravitational force plays an important role in the evolution of stars and in the behavior of galaxies. In a sense, it is gravity that holds the universe together."

—PAUL TIPLER, *PHYSICS*, 1976

MAY 27

"Newton and his theories were a step ahead of the technologies that would define his age. Thermodynamics, the grand theoretical vision of the nineteenth century, operated in the other direction with practice leading theory. The sweeping concepts of energy, heat, work and entropy, which thermodynamics (and its later form, statistical mechanics) would embrace, began first on the shop floor. Originally the domain of engineers, thermodynamics emerged from their engagement with machines. Only later did this study of heat and its transformation rise to the heights of abstract physics and, finally, to a new cosmological vision."

—ADAM FRANK, *ABOUT TIME*, 2011

MAY 28

"We have an unfortunate tendency to oversimplify invention
to lists of names, dates, and other statistics.
Look more closely and you'll find a rich and fascinating ecosystem.
It's not just the idea that counts—the way it is
implemented and the context are equally important."

—JEFF HECHT, "HOW INVENTION BEGINS, BY JOHN H. LIENHARD," *NEW SCIENTIST*, 2006

MAY 29

"What makes planets go around the sun?
At the time of Kepler, some people answered this problem by saying
that there were angels behind them beating their wings and pushing the planets
around an orbit. As you will see, the answer is not very far from the truth.
The only difference is that the angels sit in a different direction
and their wings push inward."

—RICHARD FEYNMAN, *THE CHARACTER OF PHYSICAL LAW*, 1965

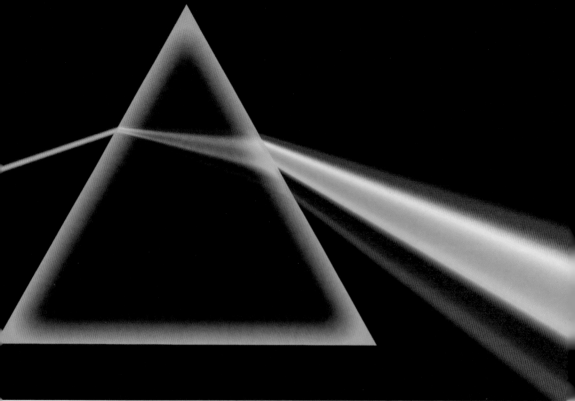

MAY 30

"Everything that has already happened is particles,
everything in the future is waves. The advancing sieve of time
coagulates waves into particles at the moment 'now.'"

—WILLIAM LAWRENCE BRAGG,
QUOTED IN RONALD CLARK'S *EINSTEIN: THE LIFE AND TIMES*, 1971

MAY 31

"There are whole walls in libraries covered by countless shelves
which are bending under the books on electronics and quantum electrodynamics.
But in none of those books will you find a proper definition of an electron,
for the very good reason that we haven't the foggiest idea what an electron 'is'."

—VINCENT ICKE, *THE FORCE OF SYMMETRY*, 1995

JUNE 1

BORN ON THIS DAY: Nicolas Léonard Sadi Carnot, 1796

"You must remember this: Like atoms, heat is so intangible that it was one of the last concepts in classical physics to be sorted out. In the process, the science of thermodynamics was created. Pollyannas who believe anything is possible should be subjected to a course in thermodynamics."

—TONY ROTHMAN, *INSTANT PHYSICS: FROM ARISTOTLE TO EINSTEIN, AND BEYOND*, 1995

JUNE 2

"Science is not about control. It is about cultivating a perpetual condition
of wonder in the face of something that forever grows one step
richer and subtler than our latest theory about it.
It is about reverence, not mastery."

—RICHARD POWERS, *THE GOLD BUG VARIATIONS*, 1991

JUNE 3

"The mathematics involved in string theory . . . in subtlety and sophistication . . .
vastly exceeds previous uses of mathematics in physical theories. . . .
String theory has led to a whole host of amazing results in mathematics
in areas that seem far removed from physics. To many this indicates
that string theory must be on the right track."

—MICHAEL ATIYAH, "PULLING THE STRINGS," *NATURE*, 2005

JUNE 4

"Consider [physicist] Andrei Linde's suggestion that, rather than there being only one universally valid set of physical laws, there are many different universes, each with its own laws of nature, each randomly different from the other. . . . Is the assumption that there is any unique universal physical law another childish dream from which we must awaken? . . . If random, [physical laws] cannot be God's thoughts, because they are not the product of any thought, much less that of God."

—PETER PESIC, "THE BELL & THE BUZZER: ON THE MEANING OF SCIENCE," *DAEDALUS*, 2003

JUNE 5

"[Isaac Newton's] theory of gravitational forces, whose main hypothesis
is the attraction of all particles of matter for one another, yields the derived law of
universal gravitation, which in turn explains, as we have seen, Kepler's empirical laws
and a wealth of other phenomena. Since one purpose of any theory is this type
of explanation and summary, Newton's theory strikes us as eminently satisfactory."

—GERALD JAMES HOLTON, *PHYSICS, THE HUMAN ADVENTURE*, 1952

JUNE 6

"Our cosmos—the world we see, hear, feel—is the three-dimensional 'surface'
of a vast, four-dimensional sea. . . . What lies outside the sea's surface?
The wholly other world of God! No longer is theology embarrassed by the
contradiction between God's immanence and transcendence.
Hyperspace touches every point of three-space. God is closer to us than our breathing.
He can see every portion of our world, touch every particle without
moving a finger though our space. Yet the Kingdom of God is completely
'outside' of three-space, in a direction in which we cannot even point."

—MARTIN GARDNER, "THE CHURCH OF THE FOURTH DIMENSION," 1962

JUNE 7

"The search [for physical laws and particles may] be over for now,
placed on hold for the next civilization with the temerity to believe that people,
pawns in the ultimate chess game, are smart enough to figure out the rules."

—GEORGE JOHNSON, "WHY IS FUNDAMENTAL PHYSICS SO MESSY?" *WIRED*, 2007

JUNE 8

"Galileo loved a fight, and he took to calling his opponents 'mental pygmies' and 'hardly deserving to be called human beings.' Two professors at his university hadn't even deigned to peer through his telescope. When one of them died a little later, Galileo wrote that he 'did not choose to see my celestial trifles while he was on earth; perhaps he will do so now that he has gone to heaven.'"

—JAMES C. DAVIS, *THE HUMAN STORY*, 2004

JUNE 9

"We now know that there exist true propositions which we can never formally prove.
What about propositions whose proofs require arguments beyond our capabilities?
What about propositions whose proofs require millions of pages?
Or a million, million pages? Are there proofs that are possible, but beyond us?"

—CALVIN CLAWSON, *MATHEMATICAL MYSTERIES*, 1996

JUNE 10

"It's a truly remarkable fact that our deepest understanding of the material world is embodied in mathematics, often in concepts that were originated with some very different motivation. A good example is our best description of how gravity works, Einstein's 1915 theory of general relativity, in which the gravitational force comes from the curvature of space and time. The formulation of this theory required Einstein to use mathematics developed 60 years earlier by the great German mathematician Bernhard Riemann, who was studying abstract questions involving geometry."

—PETER WOIT, "BOOK REVIEW: *OUR MATHEMATICAL UNIVERSE BY MAX TEGMARK,*" *WALL STREET JOURNAL,* 2014

JUNE 11

"There are many people . . . who would be perfectly able to argue the value of having read Shakespeare but would see no usefulness at all in being aware of chemical laws. . . . While it's true that such laws might not make it possible to increase your IRA earnings, they . . . describe the universe we live in and reveal the mysteries still contained in it. . . . If you are familiar with both the First and Second Law of Thermodynamics, you will be much less likely to waste money investing in a perpetual motion machine."

—JAY INGRAM, *THE BARMAID'S BRAIN AND OTHER STRANGE TALES FROM SCIENCE*, 2000

JUNE 12

"I believe that scientific knowledge has fractal properties,
that no matter how much we learn, whatever is left, however small it may seem,
is just as infinitely complex as the whole was to start with.
That, I think, is the secret of the Universe."

—ISAAC ASIMOV, *I. ASIMOV*, 1994

$$\nabla \underline{E} = \frac{1}{\varepsilon_0}$$

$$\nabla \underline{B} = 0$$

$$\nabla \times \underline{E} = \frac{\partial \underline{B}}{\partial t}$$

$$\nabla \times \underline{B} = \mu_0 \left(\underline{J} + \varepsilon_0 \frac{\partial \underline{E}}{\partial t} \right)$$

JUNE 13

BORN ON THIS DAY: Thomas Young, 1773; James Clerk Maxwell, 1831

"Although Maxwell's equations are relatively simple, they daringly reorganize our perception of nature, unifying electricity and magnetism and linking geometry, topology, and physics. They are essential to understanding the surrounding world. And as the first field equations, they not only showed scientists a new way of approaching physics but also took them on the first step towards a unification of the fundamental forces of nature."

—ROBERT P. CREASE, "THE GREATEST EQUATIONS EVER," *PHYSICS WORLD*, 2004

JUNE 14

Charles-Augustin de Coulomb, 1736

"Had [Isaac] Newton not been steeped in alchemical and other magical learning, he would never have proposed forces of attraction and repulsion between bodies as the major feature of his physical system."

—JOHN HENRY, "NEWTON, MATTER, AND MAGIC," IN *LET NEWTON BE!*, 1988

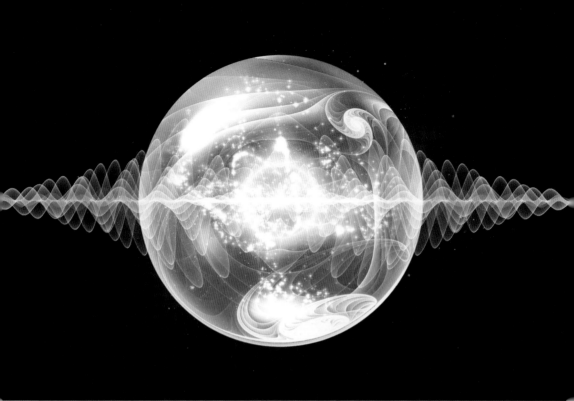

JUNE 15

"A science of all these possible kinds of space would undoubtedly be the highest
enterprise which a finite understanding could undertake in the field of geometry. . . .
If it is possible that there could be regions with other dimensions,
it is very likely that a God had somewhere brought them into being.
Such higher spaces would not belong to our world, but form separate worlds."

—IMMANUEL KANT, "THOUGHTS ON THE TRUE ESTIMATION OF LIVING FORCES," 1747

JUNE 16

"In the beginning, God said the four-dimensional divergence
of an antisymmetric, second rank tensor equals zero,
and there was Light, and it was good."

—MESSAGE ON A T-SHIRT, AS TOLD BY MICHIO KAKU,
"PARALLEL UNIVERSES, THE MATRIX, AND SUPERINTELLIGENCE," KURZWEILAI.NET, 2003

JUNE 17

"I'll be looking for you, Will, every moment, every single moment.
And when we do find each other again, we'll cling together so tight that nothing and
no one'll ever tear us apart. Every atom of me and every atom of you. . . .
We'll live in birds and flowers and dragonflies and pine trees and in clouds and in
those little specks of light you see floating in sunbeams. . . .
And when they use our atoms to make new lives, they won't just be able to take one,
they'll have to take two, one of you and one of me, we'll be joined so tight. . . ."

—PHILIP PULLMAN, *THE AMBER SPYGLASS*, 2000

JUNE 18

"Science, [Freeman] Dyson says, is an inherently subversive act. Whether overturning a longstanding idea (Heisenberg upending causality with quantum mechanics, Gödel smashing the pure platonic notion of mathematical decidability) or marshaling the same disdain for received political wisdom (Galileo, Andrei Sakharov), the scientific ethic—stubbornly following your nose where it leads you—is a threat to establishments of all kinds."

—GEORGE JOHNSON, "DANCING WITH THE STARS," *NEW YORK TIMES BOOK REVIEW*, 2007

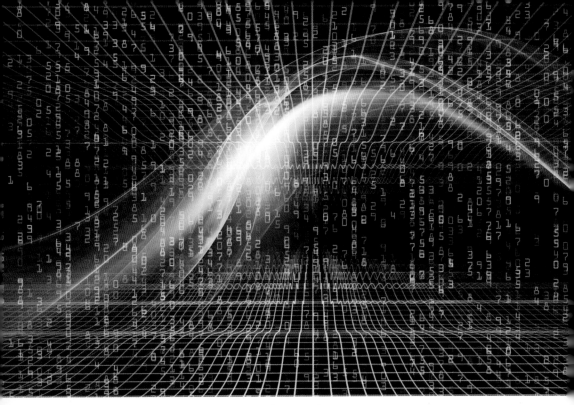

JUNE 19

"From the intrinsic evidence of his creation,
the Great Architect of the Universe
now begins to appear as a pure mathematician."

—JAMES HOPWOOD JEANS, *THE MYSTERIOUS UNIVERSE*, 1930

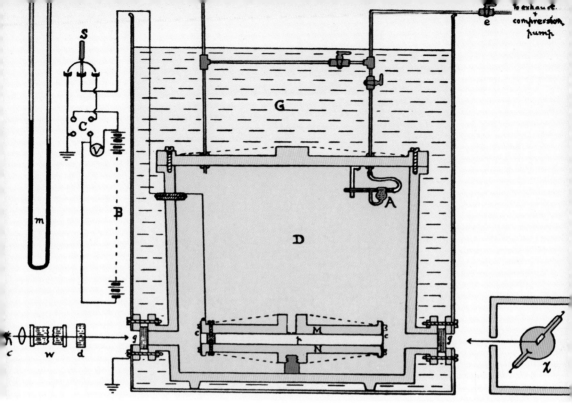

JUNE 20

"Physics makes progress because experiment constantly causes new disagreements
to break out between laws and facts, and because physicists constantly
touch up and modify laws in order that they may more faithfully represent the facts."

—PIERRE DUHEM, *THE AIM AND STRUCTURE OF PHYSICAL THEORY*, 1962

JUNE 21

"From a long view of the history of mankind—seen from, say, ten thousand years from now—there can be little doubt that the most significant event of the 19th century will be judged as Maxwell's discovery of the laws of electrodynamics. The American Civil War will pale into provincial insignificance in comparison with this important scientific event of the same decade."

–RICHARD FEYNMAN, *THE FEYNMAN LECTURES ON PHYSICS*, 1964

JUNE 22

"Great equations change the way we perceive the world. They reorchestrate the world—transforming and reintegrating our perception by redefining what belongs together with what. Light and waves. Energy and mass. Probability and position. And they do so in a way that often seems unexpected and even strange."

—ROBERT P. CREASE, "THE GREATEST EQUATIONS EVER," *PHYSICS WORLD*, 2004

JUNE 23

"There may only be a small number of laws, which are self-consistent and which lead to complicated beings like ourselves. . . . And even if there is only one unique set of possible laws, it is only a set of equations. What is it that breathes fire into the equations and makes a universe for them to govern? Is the ultimate unified theory so compelling that it brings about its own existence?"

—STEPHEN HAWKING, *BLACK HOLES AND BABY UNIVERSES AND OTHER ESSAYS*, 1993

JUNE 24

"In Newton's day . . . [some erroneous ideas concerning] the physics of motion continued to be taught from scholastic textbooks; pedantry is slow to change in any era. But by the latter part of the seventeenth century, Galileo's conception of inertia, refined and corrected, was accepted and taken for granted by most active, productive physical scientists. . . . [However] Newton set the law of inertia at the head of the laws of motion and gave it the tone of a proclamation of emancipation from scholastic theory."

—ARNOLD ARONS, *DEVELOPMENT OF CONCEPTS OF PHYSICS*, 1965

JUNE 25

"There is thus a possibility that the ancient dream of philosophers to connect all Nature with the properties of whole numbers will some day be realized. To do so physics will have to develop a long way to establish the details of how the correspondence is to be made. One hint for this development seems pretty obvious, namely, the study of whole numbers in modern mathematics is inextricably bound up with the theory of functions of a complex variable, which theory we have already seen has a good chance of forming the basis of the physics of the future. The working out of this idea would lead to a connection between atomic theory and cosmology."

—PAUL DIRAC, "THE RELATION BETWEEN MATHEMATICS AND PHYSICS,"
PROCEEDINGS OF THE ROYAL SOCIETY (EDINBURGH), 1938-1939

JUNE 26

BORN ON THIS DAY: William Thomson, Lord Kelvin, 1824

"And so in its actual procedure physics studies not these inscrutable qualities, but pointer-readings which we can observe. The readings, it is true, reflect the fluctuations of the world-qualities; but our exact knowledge is of the readings, not of the qualities. The former have as much resemblance to the latter as a telephone number has to a subscriber."

—ARTHUR STANLEY EDDINGTON, "THE DOMAIN OF PHYSICAL SCIENCE," IN JOSEPH NEEDHAM'S *SCIENCE, RELIGION AND REALITY*, 1925

JUNE 27

"Some of the scientists most closely involved, and some of the most observant philosophers of science, have taken the view that the laws of nature were: invented by man (Einstein, Bohr, Popper); not invented by man (Planck); expressions of a real underlying order in the world (Einstein); working models justified only by their utility (von Neumann, Feynman); . . . steps on the road toward complete understanding (Feynman, Deutsch); steps on a road that has no end (Born, Popper, Kuhn); . . ."

—MICHAEL FRAYN, *THE HUMAN TOUCH*, 2007

JUNE 28

"The law that entropy increases—the Second Law of Thermodynamics—holds, I think, the supreme position among the laws of Nature. If someone points out to you that your pet theory of the Universe is in disagreement with Maxwell's equations— then so much the worse for Maxwell's equations. If it is found to be contradicted by observation—well, these experimentalists do bungle things sometimes. But if your theory is found to be against the Second Law of Thermodynamics I can give you no hope; there is nothing for it but to collapse in deepest humiliation."

—ARTHUR STANLEY EDDINGTON, *THE NATURE OF THE PHYSICAL WORLD*, 1928

JUNE 29

"Our brains have evolved to get us out of the rain, find where the berries are, and keep us from getting killed. Our brains did not evolve to help us grasp really large numbers or to look at things in a hundred thousand dimensions."

—RONALD GRAHAM, QUOTED IN PAUL HOFFMAN'S
"THE MAN WHO LOVES ONLY NUMBERS," *ATLANTIC MONTHLY*, 1987

JUNE 30

"Until now, physical theories have been regarded as merely models
which approximately describe the reality of nature. As the models improve,
so the fit between theory and reality gets closer. Some physicists are
now claiming that supergravity is the reality, that the model and the real world
are in mathematically perfect accord."

—PAUL DAVIES, *SUPERFORCE*, 1984

185

JULY 1

"Admit the existence of a God, of a personal God,
and the possibility of miracle follows at once. If the laws of nature are carried on
in accordance with His will, He who willed them may will their suspension.
And if any difficulty should be felt as to their suspension,
we are not even obliged to suppose that they have been suspended."

—GEORGE GABRIEL STOKES, *NATURAL THEOLOGY*, 1891

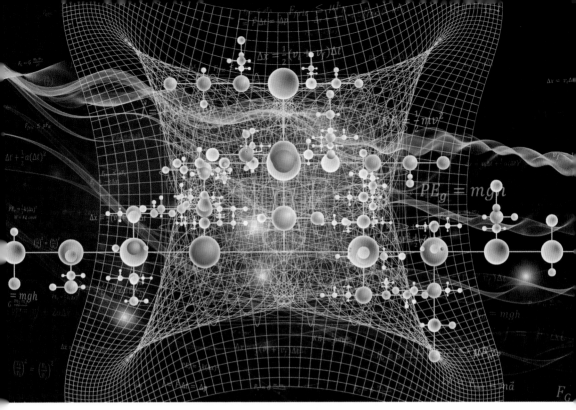

JULY 2

"If tachyons are one day discovered, . . . the day before that
momentous occasion a notice from the discovers should appear in newspapers,
announcing 'Tachyons have been discovered tomorrow.'"

—PAUL NAHIN, *TIME MACHINES*, 1993

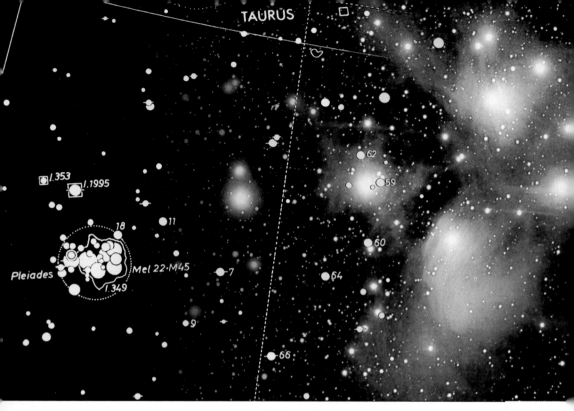

JULY 3

"Looking at the stars always makes me dream, as simply as I dream
over the black dots representing towns and villages on a map.
Why, I ask myself, shouldn't the shining dots of the sky be as accessible
as the black dots on the map of France? If we take the train to get to Tarascon
or Rouen, we take death to reach a star. One thing undoubtedly true
in this reasoning is this: that while we are alive we cannot get to a star,
any more than when we are dead we can take the train."

—VINCENT VAN GOGH, LETTER TO THEO VAN GOGH, JULY 1888

JULY 4

"Some men still believe in the mathematical design of nature.
They may grant that many of the earlier mathematical theories of physical phenomena
were imperfect, but they point to continuing improvements that not only embrace
more phenomena but offer far more accurate agreement with observations.
Thus Newtonian mechanics replaced Aristotelian mechanics, and the theory of
relativity improved on Newtonian mechanics. Does not this history imply that there is
design and that man is approaching closer and closer to the truth?"

—MORRIS KLINE, *MATHEMATICS: THE LOSS OF CERTAINTY*, 1980

JULY 5

"As pilgrimages to the shrines of saints draw thousands of English Catholics to the Continent, there may be some persons in the British Islands sufficiently in love with science, not only to revere the memory of its founders, but to wish for a description of the locality and birth-place of a great master of knowledge—John Dalton [developer of modern atomic theory]—who did more for the world's civilisation than all the reputed saints in Christendom."

—HENRY LONSDALE, *THE WORTHIES OF CUMBERLAND: JOHN DALTON*, 1874

JULY 6

"Science cannot solve the ultimate mystery of nature.
And that is because, in the last analysis, we ourselves are part of nature
and therefore part of the mystery that we are trying to solve."

—MAX PLANCK, *WHERE IS SCIENCE GOING?*, 1933

JULY 7

"Maybe the brilliance of the brilliant can be understood only by the nearly brilliant."

—ANTHONY SMITH, *THE MIND*, 1984

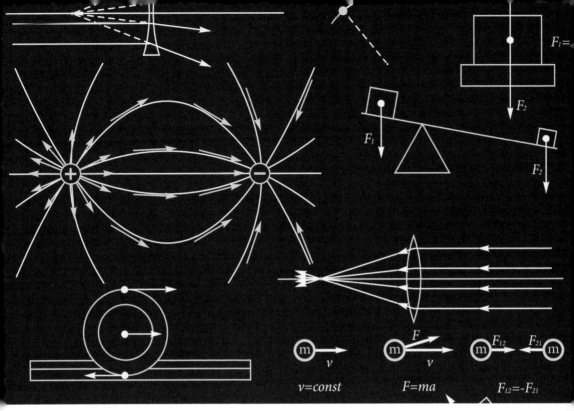

JULY 8

"At every major step physics has required, and frequently stimulated,
the introduction of new mathematical tools and concepts. Our present understanding
of the laws of physics, with their extreme precision and universality,
is only possible in mathematical terms."

—MICHAEL ATIYAH, "PULLING THE STRINGS," *NATURE*, 2005

JULY 9

BORN ON THIS DAY: John Wheeler, 1911

"Since 1960 the universe has taken on a wholly new face.
It has become more exciting, more mysterious, more violent, and more extreme
as our knowledge concerning it has suddenly expanded. And the most exciting,
most mysterious, most violent, and most extreme phenomenon of all has the simplest,
plainest, calmest, and mildest name—nothing more than a black hole."

—ISAAC ASIMOV, *THE COLLAPSING UNIVERSE*, 1977

JULY 10

"If we go back to our checker game, the fundamental laws are rules
by which the checkers move. Mathematics may be applied in the complex situation
to figure out what in given circumstances is a good move to make.
But very little mathematics is needed for the simple fundamental character
of the basic laws. They can be simply stated in English for checkers."

—RICHARD FEYNMAN, *THE CHARACTER OF PHYSICAL LAW*, 1965

JULY 11

"God, we are told, is not a puppet-master in regard either to human actions or to the processes of the world. If we are to exist in an environment where we can live lives of productive work and consistent understanding . . . the world has to have a regular order and pattern of its own. Effects follow causes in a way that we can chart, and so can make some attempt at coping with. So there is something odd about expecting that God will constantly step in if things are getting dangerous."

—ROWAN WILLIAMS, "OF COURSE THIS MAKES US DOUBT GOD'S EXISTENCE," SUNDAY *TELEGRAPH*, 2005

JULY 12

"Historically there was a change in the nature of scientific models with Newton. Ptolemy, Copernicus, and Newton all developed models that correctly predicted planetary motion. While it is sometimes claimed that the Ptolemaic and Copernican models were only descriptive in contrast to Newton's which was explanatory, it is more precise to say that Newton introduced a higher level of abstraction using ideas farther removed from the observations."

—BYRON JENNINGS, "ON THE NATURE OF SCIENCE," *LA PHYSIQUE AU CANADA*, 2007

JULY 13

"[Isaac Newton] was born into a world of darkness, obscurity, and magic . . .
veered at least once to the brink of madness . . . and yet discovered more of the
essential core of human knowledge than anyone before or after.
He was chief architect of the modern world. . . . He made knowledge
a thing of substance: quantitative and exact. He established principles,
and they are called his laws."

—JAMES GLEICK, *ISAAC NEWTON*, 2003

JULY 14

"The laws of nature are but the transcript of the thoughts of God,
immutable and unchangeable. There is no such thing as Chance. Chance has no
existence under the constant laws of nature or under any laws. What is called Chance
is only the uncalculated result of some known or unknown law of nature."

—HENRY AUGUSTUS MOTT, *THE LAWS OF NATURE
AND MAN'S POWER TO MAKE THEM SUBSERVIENT TO HIS WISHES*, 1882

JULY 15

"Whenever a theory appears to you as the only possible one,
take this as a sign that you have neither understood the theory nor the problem
which it was intended to solve."

—KARL POPPER, *OBJECTIVE KNOWLEDGE: AN EVOLUTIONARY APPROACH*, 1972

JULY 16

"Physics isn't Christian, though it was invented by Christians.
Algebra isn't Muslim, even though it was invented by Muslims.
Whenever we get at the truth, we transcend culture, we transcend our upbringing.
The discourse of science is a good example of where we should hold out hope
for transcending our tribalism."

—SAM HARRIS, "THE GOD DEBATE," *NEWSWEEK*, 2007

JULY 17

BORN ON THIS DAY: Georges Lemaître, 1894

"I'll tell you what the big bang was, Lestat.
It was when the cells of God began to divide."

—ANNE RICE, *TALE OF THE BODY THIEF*, 1992

JULY 18

BORN ON THIS DAY: Hendrik Lorentz, 1853

"What makes the theory of relativity so acceptable to physicists in spite of its going against the principle of simplicity is its great mathematical beauty. This is a quality which cannot be defined, any more than beauty in art can be defined, but which people who study mathematics usually have no difficulty in appreciating. . . . The restricted theory changed our ideas of space and time in a way that may be summarised by stating that the group of transformations to which the space-time continuum is subject must be changed from the Galilean group to the Lorentz group."

—PAUL DIRAC, "THE RELATION BETWEEN MATHEMATICS AND PHYSICS,"
PROCEEDINGS OF THE ROYAL SOCIETY (EDINBURGH), 1938-1939

JULY 19

"Without all doubt, this world, so diversified with that variety of forms and motions we find in it, could arise from nothing but the perfectly free will of God directing and presiding over all. From this fountain . . . the laws of Nature have flowed, in which there appear many traces indeed of the most wise contrivance, but not the least shadow of necessity. These, therefore, we must not seek from uncertain conjectures, but learn them from observations and experiments."

—ROGER COTES, PREFACE TO THE SECOND EDITION OF NEWTON'S *PRINCIPIA*, 1713

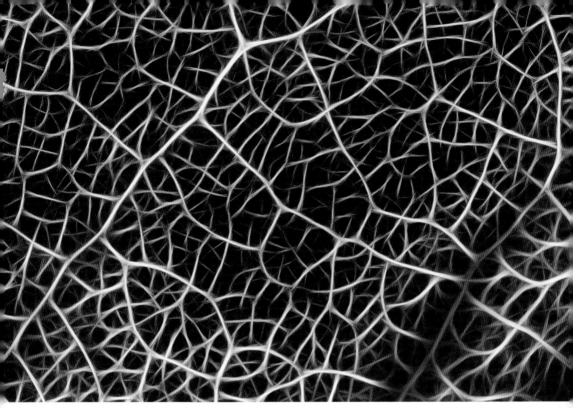

JULY 20

"We must be clear that when it comes to atoms, language can be used only as in poetry. The poet, too, is not nearly so concerned with describing facts as with creating images and establishing mental connections."

—NIELS BOHR, QUOTED BY WERNER HEISENBERG IN *PHYSICS AND BEYOND*, 1971

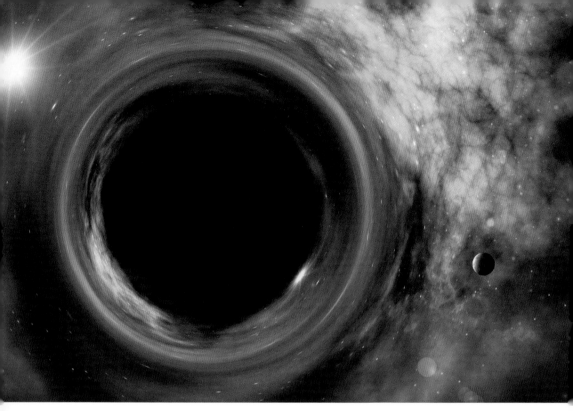

JULY 21

"We showed that if general relativity is correct, any reasonable model of the universe must start with a singularity. . . . I now think that although there is a singularity, the laws of physics can still determine how the universe began."

—STEPHEN HAWKING, *BLACK HOLES AND BABY UNIVERSES AND OTHER ESSAYS*, 1993

JULY 22

"The future religion of humanity will be based on scientific laws."

—GREG WHITEFIELD, IN POST B. BASNET'S "NEPAL BECOMING MECCA FOR BUDDHIST STUDIES," KANTIPURONLINE.COM, 2005

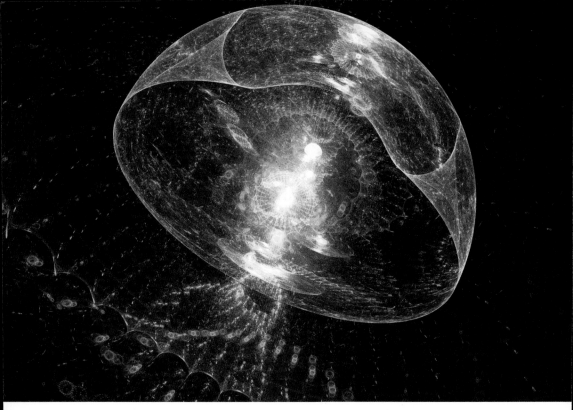

JULY 23

"When we look at the glory of stars and galaxies in the sky and the glory of forests
and flowers in the living world around us, it is evident that God loves diversity.
Perhaps the universe is constructed according to a principle of maximum diversity.
[That principle] says that the laws of nature . . . are such as to make the universe
as interesting as possible. As a result, life is possible but not too easy.
Maximum diversity often leads to maximum stress. In the end we survive,
but only by the skin of our teeth."

—FREEMAN DYSON, "NEW MERCIES: THE PRICE AND PROMISE OF HUMAN PROGRESS," *SCIENCE & SPIRIT*, 2000

JULY 24

"[Isaac] Newton was the one man who was in equal measure a creative mathematician and a creative physicist. He was one of the few physicists equally adept at theory and at experiment. His invention of the reflecting telescope, much less celestial mechanics, would have ensured him a prominent place in the history of astronomy. Newton's contemporaries remarked upon his extraordinary intuition. He seemed to know things that even he could not prove."

—ERNEST ABERS AND CHARLES F. KENNEL, *MATTER IN MOTION*, 1977

JULY 25

"If the cosmos were suddenly frozen, and all movement ceased, a survey of its structure would not reveal a random distribution of parts. Simple geometrical patterns, for example, would be found in profusion—from the spirals of galaxies to the hexagonal shapes of snow crystals. Set the clockwork going, and its parts move rhythmically to laws that often can be expressed by equations of surprising simplicity. And there is no logical or a priori reason why these things should be so."

—MARTIN GARDNER, "ORDER AND SURPRISE," *PHILOSOPHY OF SCIENCE*, 1950

JULY 26

"Since Galileo's time, science has become steadily more mathematical. . . .
It is virtually an article of faith for most theoreticians . . . that there exists a
fundamental equation to describe the phenomenon they are studying. . . .
Yet . . . it may eventually turn out that fundamental laws of nature do not need
to be stated mathematically and that they are better expressed in other ways,
like the rules governing the game of chess."

—GRAHAM FARMELO, FOREWORD TO *IT MUST BE BEAUTIFUL*, 2003

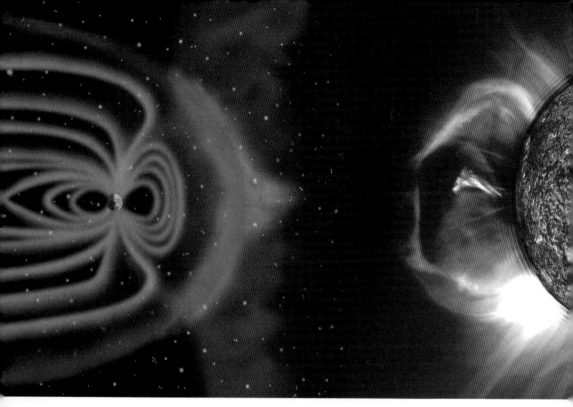

JULY 27

"The job of science is to enable the inquiring mind
to feel at home in a mysterious universe."

—LEWIS CARROLL EPSTEIN, *RELATIVITY VISUALIZED*, 1984

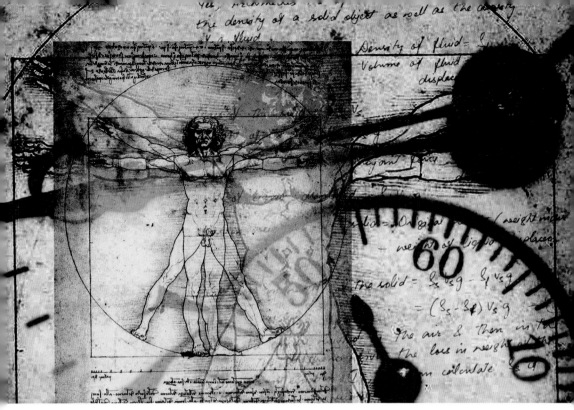

JULY 28

BORN ON THIS DAY: Robert Hooke, 1635; Charles Hard Townes, 1915

"Robert Hooke is one of the most neglected natural philosophers of all time.
The inventor of . . . the iris diaphragm in cameras, the universal joint
used in motor vehicles, the balance wheel in a watch, the originator
of the word 'cell' in biology, he was . . . architect, experimenter,
[and astronomer]—yet is known mostly for Hooke's Law [of elasticity]. . . .
He was Europe's last Renaissance man, and England's Leonardo."

—ROBERT HOOKE SCIENCE CENTRE, WWW.ROBERTHOOKE.ORG.UK, 2007

JULY 29

"The progress of mathematics and physics impels us to fly away on the wings of the poetic imagination out beyond the frontiers of Euclidean space, and to attempt to conceive of space in which more than three coordinates can stand perpendicularly to one another. But all such endeavors to fly out beyond our frontiers always end with our falling back with singed wings on the ground of our Euclidean three-dimensional space. . . . We can certainly calculate with [higher-dimensional spaces]. But we cannot conceive of them. We are confined within the space in which we find ourselves when we enter into our existence, as though in a prison. Two-dimensional beings can believe in a third dimension. But they cannot see it."

—KARL HEIM, *CHRISTIAN FAITH AND NATURAL SCIENCE*, 1953

JULY 30

"While Archimedes was turning the problem over, he chanced to come to the place of bathing, and there, as he was sitting down in the tub, he noticed that the amount of water which flowed over the tub was equal to the amount by which his body was immersed. This showed him a means of solving the problem. . . . In his joy, he leapt out of the tub and, rushing naked towards his home, he cried out with a loud voice that he had found what he sought."

—MARCUS VITRUVIUS, *DE ARCHITECTURA (ON ARCHITECTURE)*, C. 15 BCE

JULY 31

"The core of science is not a mathematical model; it is intellectual honesty."

—SAM HARRIS, "BEYOND BELIEF: SCIENCE, RELIGION, REASON, AND SURVIVAL,"
MEETING AT THE SALK INSTITUTE FOR BIOLOGICAL STUDIES, LA JOLLA, CALIFORNIA, 2006

AUGUST 1

"People say to me, 'Are you looking for the ultimate laws of physics?' No, I'm not; I'm just looking to find out more about the world and if it turns out there is a simple ultimate law which explains everything, so be it; that would be very nice to discover. If it turns out it's like an onion with millions of layers, and we're just sick and tired of looking at the layers, then that's the way it is . . ."

—RICHARD FEYNMAN, FROM AN INTERVIEW FOR THE BBC TELEVISION PROGRAM *HORIZON*, 1981

AUGUST 2

"Entropy is one of those words that almost everyone has heard and almost nobody can really explain. . . . [It's the] amount of disorder and information in a system. [Consider] a fresh, unshuffled deck of cards. In that state it has low entropy and contains little information. Just two pieces of data (the hierarchy of suits and the relative ranks of the cards) tell you where to find every card in the deck without looking. [After shuffling,] the deck has a lot of entropy and a lot of information. If you want to locate a particular card, you have to hunt through the entire deck. There is only one perfectly ordered state but about 10^{68} disordered ones, which is why you will never, ever accidentally shuffle the deck back into its original order."

—COREY S. POWELL, "WELCOME TO THE MACHINE," *NEW YORK TIMES*, 2006

AUGUST 3

"Astrophysicists have the formidable privilege of having the largest view of the Universe; particle detectors and large telescopes are today used to study distant stars, and throughout space and time, from the infinitely large to the infinitely small, the Universe never ceases to surprise us by revealing its structures little by little."

—JEAN-PIERRE LUMINET, *BLACK HOLES*, 1992

AUGUST 4

"Just as a great river is fed by small streams, some even barely noticeable . . . ,
so science and technology proceeds from small individual contributions until
it becomes an ever-increasing flow of knowledge and techniques.
This big river of fluid mechanics is closely associated with Daniel Bernoulli,
the author of the first textbook in this field."

—G. A. TOKATY, *A HISTORY AND PHILOSOPHY OF FLUID MECHANICS*, 1971

AUGUST 5

"The idea of eternally true laws of nature is a beautiful vision, but is it really an escape from philosophy and theology? For, as philosophers have argued, we can test the predictions of a law of nature and see if they are verified or contradicted, but we can never prove a law must always be true. So if we believe a law of nature is eternally true, we are believing in something that logic and evidence cannot establish."

—LEE SMOLIN, "NEVER SAY ALWAYS," *NEW SCIENTIST*, 2006

AUGUST 6

"Models in the mathematical, physical, and mechanical sciences
are of the greatest importance. Long ago philosophy perceived the essence of our
process of thought to lie in the fact that we attach to the various real objects
around us particular physical attributes—our concepts—and by means of these try
to represent the objects to our minds. . . . On this view our thoughts stand to things
in the same relation as models to the objects they represent."

—LUDWIG BOLTZMANN, *ENCYCLOPAEDIA BRITANNICA*, 1902

AUGUST 7

"The secularization of the concept of nature's laws proceeded more slowly in England than on the continent of Europe. By the end of the eighteenth century, after the French Revolution, Laplace could boast that he had no need of the 'hypothesis' of God's existence, and Kant had sought to ground the universality and necessity of Newton's laws not in God or nature, but in the constitution of human reason. . . . [Nonetheless] whether the laws of nature might be expressions of divine will was still much debated in the third quarter of the nineteenth century in Britain. . . . Not until Darwin's revolution had worked its way through British intellectual life did the laws of nature get effectively separated from God's will."

—RONALD N. GIERE, *SCIENCE WITHOUT LAWS*, 1999

AUGUST 8

BORN ON THIS DAY: Paul Dirac, 1902

"Starting in the 1890s, a series of astonishing experimental discoveries and theoretical concepts started to open up the inside of the atom to the figurative gaze of the wide-eyed physicists. . . . The key to the advances of what the late cosmologist George Gamow described as 'thirty years that shook physics' was the development of what we term the quantum theory of radiation—the idea that electromagnetic energy, light, and other forms of radiation come in the form of discrete packets . . ."

—HENRY HOOPER AND PETER GWYNNE, *PHYSICS AND THE PHYSICAL PERSPECTIVE*, 1980

AUGUST 9

"We live on an island of knowledge surrounded by a sea of ignorance.
As our island of knowledge grows, so does the shore of our ignorance."

—JOHN A. WHEELER, QUOTED IN JOHN HORGAN'S "THE NEW CHALLENGES,"
SCIENTIFIC AMERICAN, 1992

AUGUST 10

"There is no mention of laws in Copernicus, or even in Galileo,
while Kepler makes no use of the term himself in introducing what are often described
as the first truly scientific laws—his three laws of planetary motion."

—MICHAEL FRAYN, *THE HUMAN TOUCH*, 2007

AUGUST 11

"Newton was probably responsible for the concept that there are
seven primary colours in the spectrum—he had a strong interest in musical harmonies
and, since there are seven distinct notes in the musical scale, he divided up the
spectrum into spectral bands with widths corresponding to the ratios of the small
whole numbers found in the just scale."

—MALCOLM LONGAIR, "LIGHT AND COLOUR," IN *COLOUR: ART & SCIENCE*, 1995

AUGUST 12

BORN ON THIS DAY: Erwin Schrödinger, 1887

"Many classic physics equations—including $E = mc^2$ and Schrödinger's equation—were not conclusions drawn from statements about observations. Rather, they were conclusions based on reasoning from other equations and information; they are therefore more like theorems. And theorems can be equation-like for their strong empirical content and value."

—ROBERT P. CREASE, "THE GREATEST EQUATIONS EVER," *PHYSICS WORLD*, 2004

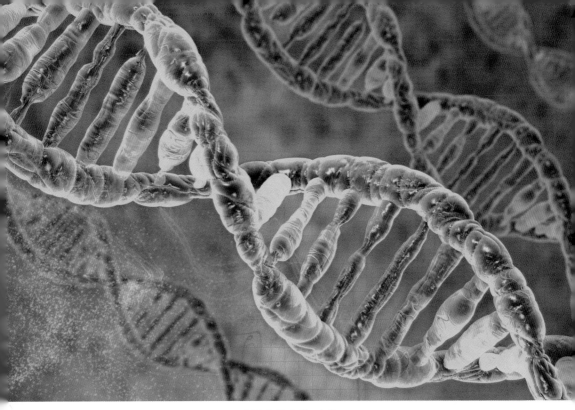

AUGUST 13

"The laws of the universe are cunningly contrived to coax life into being . . .
If life follows from [primordial] soup with causal dependability, the laws of nature
encode a hidden subtext . . . which tells them: 'Make life!' . . . It means that
the laws of the universe have engineered their own comprehension."

—PAUL DAVIES, *THE FIFTH MIRACLE*, 2000

AUGUST 14

BORN ON THIS DAY: Hans Christian Ørsted, 1777

"Together with [Hans] Ørsted, [Michael Faraday] had shown that electricity could beget magnetism, and magnetism could beget electricity, a genetic relationship so incestuous and circular there was none other like it in Nature. . . . The son of a common laborer had discerned and written down a great secret of the natural world, one that would spell the end of the Industrial Revolution and the beginning of the Electrical Age."

—MICHAEL GUILLEN, *FIVE EQUATIONS THAT CHANGED THE WORLD*, 1995

AUGUST 15

BORN ON THIS DAY: Louis de Broglie, 1892

"'And yet what can there be,' we continue to ask perversely, 'beyond the quantum mechanical wave function that may someday be written down to describe a multiverse in which the electron takes every possible path?' Newton's laboratory table, perhaps, on which our multiverse sits enclosed in a crystalline sphere, dreaming that it is everything."

—GEORGE ZEBROWSKI, "TIME IS NOTHING BUT A CLOCK," *OMNI*, 1994

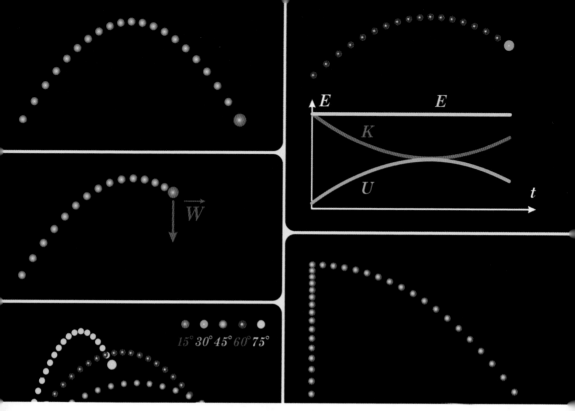

AUGUST 16

"This law [of gravitation] has been called 'the greatest generalization achieved by the human mind'. . . . I am interested not so much in the human mind as in the marvel of a nature which can obey such an elegant and simple law as this law of gravitation. Therefore our main concentration will not be on how clever we are to have found it all out, but on how clever nature is to pay attention to it."

—RICHARD FEYNMAN, *THE CHARACTER OF PHYSICAL LAW*, 1965

AUGUST 17

"This property of human languages—their resistance to algorithmic processing—
is perhaps the ultimate reason why only mathematics can furnish an adequate
language for physics. It is not that we lack words for expressing all this $E = mc^2$ and
$\int e^{iS(\phi)} D\phi$ stuff . . . , the point is that we still would not be able to do anything
with these great discoveries if we had only words for them. . . . Miraculously,
it turns out that even very high level abstractions can somehow reflect reality:
knowledge of the world discovered by physicists can be expressed
only in the language of mathematics."

—YURI I. MANIN, "MATHEMATICAL KNOWLEDGE: INTERNAL, SOCIAL, AND CULTURAL ASPECTS,"
MATHEMATICS AS METAPHOR: SELECTED ESSAYS, 2007

AUGUST 18

"Somewhere in that great ocean of truth, the answers to questions about life in the
universe are hidden. . . . Beyond these questions are others that we cannot even ask,
questions about the universe as it may be perceived in the future
by minds whose thoughts and feelings are as inaccessible to us as our
thoughts and feelings are inaccessible to earthworms."

—FREEMAN DYSON, "SCIENCE & RELIGION: NO ENDS IN SIGHT," *NEW YORK REVIEW OF BOOKS*, 2002

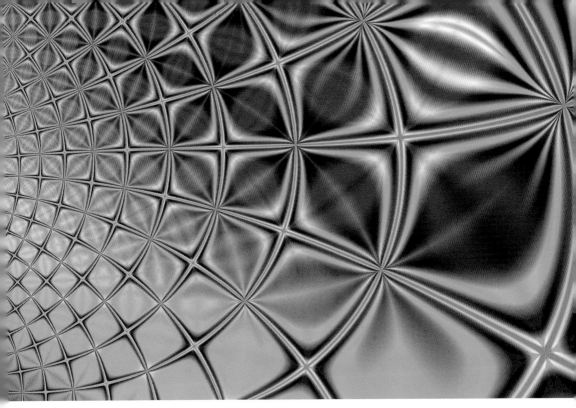

AUGUST 19

"We have drawn a picture to represent the fact that people are persistent spacetime patterns. . . . The simple processes of eating and breathing weave all of us together into a vast four-dimensional tapestry. No matter how isolated you may sometimes feel, no matter how lonely, you are never really cut off from the whole."

—RUDY RUCKER, *THE FOURTH DIMENSION*, 1985

AUGUST 20

"Modern science is a newcomer, barely four hundred years old.
Though indebted in deep ways to Plato, Aristotle, and Greek natural philosophy,
the pioneers of the 'new philosophy' called for a decisive break
with ancient authority. In 1536, Pierre de La Ramée defended the provocative
thesis that 'everything Aristotle said is wrong.'"

—PETER PESIC, "THE BELL & THE BUZZER: ON THE MEANING OF SCIENCE," *DAEDALUS*, 2003

AUGUST 21

"I am convinced more and more that the necessary truth of our geometry cannot be demonstrated, at least not by the human intellect to the human understanding. Perhaps in another world, we may gain other insights into the nature of space which at present are unattainable to us. Until then we must consider geometry as of equal rank not with arithmetic, which is purely a priori, but with mechanics."

—CARL GAUSS, LETTER TO HEINRICH WILHELM OLBERS, 1817

AUGUST 22

"How often might a man, after he had jumbled a set of letters in a bag, fling them out upon the ground before they would . . . make a good discourse in prose? And may not a little book be as easily made by chance, as this great volume of the world? How long might a man be in sprinkling colours upon canvas with a careless hand, before they would happen to make the exact picture of a man? And is a man easier made by chance than his picture? How long might twenty thousand blind men . . . wander up and down before they would all meet upon Salisbury Plains, and fall into rank and file in the exact order of an army? And yet this is much more easy to be imagined than how the innumerable blind parts of matter should rendezvous themselves into a world."

—JOHN TILLOTSON, *MAXIMS AND DISCOURSES, MORAL AND DIVINE*, 1719

AUGUST 23

"We are in the habit of talking as if [the laws of Nature] caused events to happen; but they have never caused any event at all. The laws of motion do not set billiard balls moving; they analyse the motions after something else . . . has provided it. They produce no events: they state the pattern to which every event . . . must conform. . . . It is therefore inaccurate to define a miracle as something that breaks the laws of Nature. . . . If God creates a miraculous spermatozoon in the body of a virgin, it does not proceed to break any laws. . . . Nature is ready. Pregnancy follows, according to all the normal laws, . . ."

—C. S. LEWIS, MIRACLES, REPRINTED IN *THE COMPLETE C. S. LEWIS SIGNATURE CLASSICS*, 2002

AUGUST 24

"No theory can be objective, actually coinciding with nature. . . . rather . . . each theory is only a mental picture of phenomena, related to them as sign is to designation. . . . From this it follows that it cannot be our task to find an absolutely correct theory but rather a picture that is as simple as possible and which represents phenomena as accurately as possible. One might even conceive of two quite different theories both equally simple and equally congruent with phenomena, which therefore, in spite of their difference, are equally correct."

—LUDWIG BOLTZMANN, "ON THE DEVELOPMENT OF METHODS OF THEORETICAL PHYSICS IN RECENT TIMES," 1905

AUGUST 25

"Mathematicians define things, otherwise they wouldn't have a clue what they are talking about. This is so because everything in mathematics was invented by people. Contrariwise, nothing of the substance of physics was invented by us: Nature is out yonder and there is not a shred of evidence that the Universe cares a fig about human or other beings. The formulations of physics are uniquely ours, but those, of course, are mathematics. That is why, contrary to popular opinion, physicists never actually define anything physical."

—VINCENT ICKE, *THE FORCE OF SYMMETRY*, 1995

AUGUST 26

"It doesn't make much difference whether this determinism is due to
an omnipotent God or to the laws of science. Indeed, one could always say
that the laws of science are the expression of the will of God."

—STEPHEN HAWKING, *BLACK HOLES AND BABY UNIVERSES AND OTHER ESSAYS*, 1993

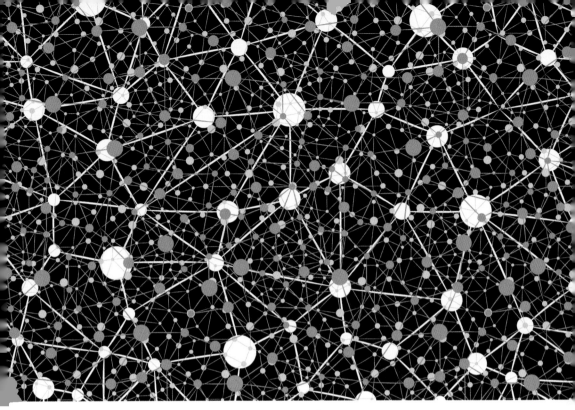

AUGUST 27

"Suppose we divide the space into little volume elements.
If we have black and white molecules, how many ways could we distribute them
among the volume elements so that white is on one side and black is on the other?
On the other hand, how many ways could we distribute them with no restriction on
which goes where? Clearly, there are many more ways to arrange them
in the latter case. We measure 'disorder' by the number of ways that the insides can be
arranged, so that from the outside it looks the same. The logarithm of that number of
ways is the entropy. The number of ways in the separated case is less,
so the entropy is less, or the 'disorder' is less."

—RICHARD FEYNMAN, "ORDER AND ENTROPY," *THE FEYMAN LECTURES ON PHYSICS*, 1964

AUGUST 28

"In science we aim for a picture of nature as it really is, unencumbered by any philosophical or theological prejudice. Some see the search for scientific truth as a search for an unchanging reality behind the ever-changing spectacle we observe with our senses. The ultimate prize in that search would be to grasp a law of nature— a part of a transcendent reality that governs all change, but itself never changes."

—LEE SMOLIN, "NEVER SAY ALWAYS," *NEW SCIENTIST*, 2006

AUGUST 29

"Our entire universe may slowly stop expanding, go into a contracting phase,
and finally disappear into a black hole,
like an acrobatic elephant jumping into its anus."

—MARTIN GARDNER, "SEVEN BOOKS ON BLACK HOLES," *SCIENCE: GOOD, BAD, AND BOGUS*, 1981

AUGUST 30

BORN ON THIS DAY: Ernest Rutherford, 1871

"Rutherford was as straightforward and unpretentious as a physicist as he was
elsewhere in life, and that no doubt was one of the secrets of his success.
'I was always a believer in simplicity, being a simple man myself,' he said.
If a principle of physics could not be explained to a barmaid, he insisted,
the problem was with the principle, not the barmaid."

—WILLIAM H. CROPPER, *GREAT PHYSICISTS*, 2004

AUGUST 31

BORN ON THIS DAY: Hermann Ludwig Ferdinand von Helmholtz, 1821

"Change one event in the past and you get a brand new future?
Erase the conquest of Alexander by nudging a Neolithic pebble? Extirpate America by
pulling up a shoot of Sumerian grain? Brother, that isn't the way it works at all!
The space-time continuum's built of stubborn stuff and change is anything but a
chain-reaction. Change the past and you start a wave of changes moving futurewards,
but it damps out mighty fast. Haven't you ever heard of temporal reluctance,
or of the Law of the Conservation of Reality?"

—FRITZ LEIBER, "TRY AND CHANGE THE PAST," *ASTOUNDING SCIENCE FICTION*, 1958

SEPTEMBER 1

"Newton's aim was to unravel nothing less than God's secret messages. . . .
Above all Newton was intent on finding out when the world would come to an end.
Then, he believed, Christ would return and set up a 1,000-year Kingdom of God
on earth and he—Isaac Newton, that is—would rule the world as one among
the saints. . . . [Newton] had calculated the year of the apocalypse: 2060."

—GEORGE G. SZPIRO, *THE SECRET LIFE OF NUMBERS*, 2006

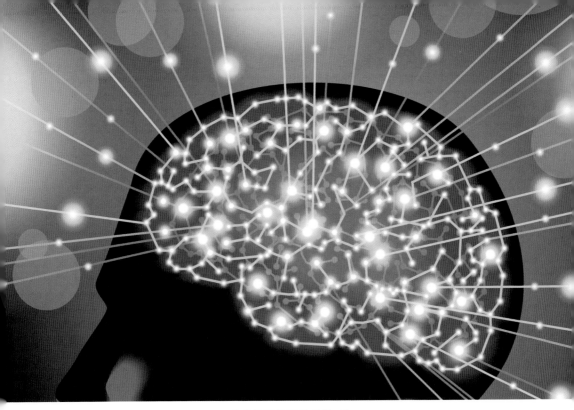

SEPTEMBER 2

"When the scientist attempts to understand a group of natural phenomena,
he begins with the assumption that these phenomena obey certain laws which,
being intelligible to our reason, can be comprehended. This is not,
let us hasten to note, a self-evident postulate which leaves no room for qualifications.
In effect, what it does is to reiterate the rationality of the physical world,
to recognize that the structure of the material universe has something in common
with the laws that govern the behavior of the human mind."

—ARTHUR MARCH AND IRA M. FREEMAN, *THE NEW WORLD OF PHYSICS*, 1962

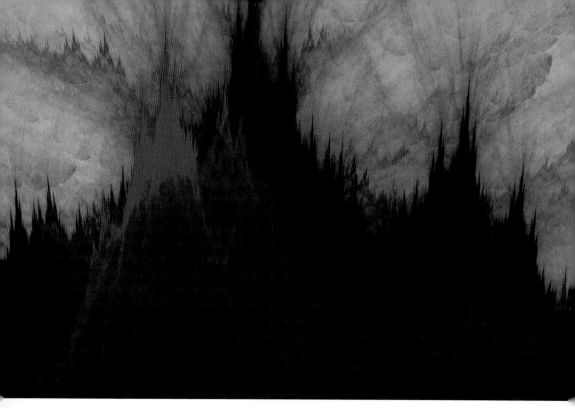

SEPTEMBER 3

"[Hermann] Weyl said that God exists since mathematics is consistent and the devil exists since we cannot prove the consistency."

—MORRIS KLINE, *MATHEMATICAL THOUGHT FROM ANCIENT TO MODERN TIMES*, 1990

SEPTEMBER 4

"Without regularities embodied in the laws of physics,
we would be unable to make sense of physical events; without regularities in
the laws of nature, we would be unable to discover the laws themselves."

—GERD BAUMANN, *SYMMETRY ANALYSIS OF DIFFERENTIAL EQUATIONS WITH MATHEMATICA*, 2000

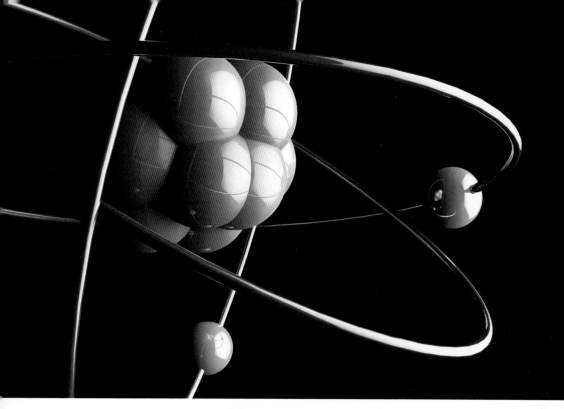

SEPTEMBER 5

"Only those quantities that can be measured have any real meaning in physics. If we could focus a 'super' microscope on an electron in an atom and see it moving around in an orbit, we would declare that such orbits have meaning. However, we shall show that it is fundamentally impossible to make such an observation—even with the most ideal instrument that could conceivably be constructed. Therefore, we declare that such orbits have no physical meaning."

—DAVID HALLIDAY AND ROBERT RESNICK, *PHYSICS*, 1966

SEPTEMBER 6

"Imagine the chaos that would arise if time machines were as common as automobiles, with tens of millions of them commercially available. Havoc would soon break loose, tearing at the fabric of our universe. Millions of people would go back in time to meddle with their own past and the past of others, rewriting history in the process. . . . It would thus be impossible to take a simple census to see how many people there were at any given time."

—MICHIO KAKU, *HYPERSPACE*, 1995

SEPTEMBER 7

"I wanted to become a theologian. For a long time I was restless.
Now, however, behold how through my effort God is being celebrated in astronomy."

—JOHANNES KEPLER, LETTER TO MICHAEL MAESTLIN, 1595

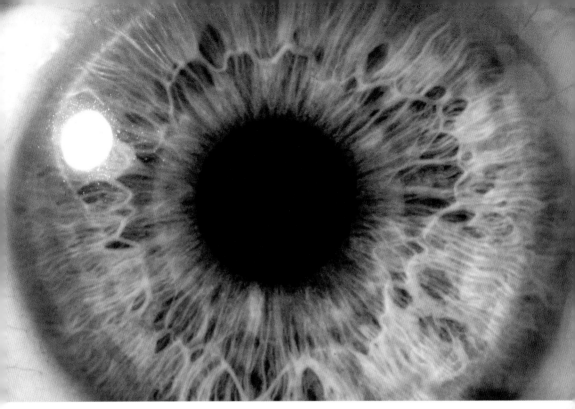

SEPTEMBER 8

"[Isaac] Newton was a decidedly odd figure . . . famously distracted
(upon swinging his feet out of bed in the morning he would reportedly
sometimes sit for hours, immobilized by the sudden rush of thoughts to his head),
and capable of the most riveting strangeness. . . . Once he inserted a bodkin [a long
needle for sewing leather] into his eye socket . . . just to see what would happen."

—BILL BRYSON, *A SHORT HISTORY OF NEARLY EVERYTHING*, 2004

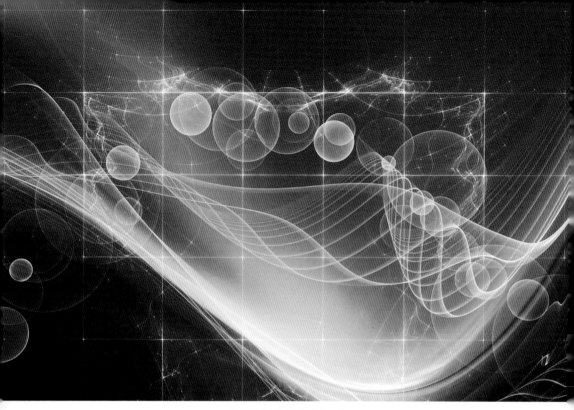

SEPTEMBER 9

"The universe seems to operate by several sets of rules that act in layers,
independently of each other. The most apparent of these basic rules of nature, gravity,
controls the biggest objects in the universe: the stars, the planets, you and me.
The other three that scientists have uncovered operate at the subatomic level."

—JOHN BOSLOUGH, *STEPHEN HAWKING'S UNIVERSE*, 1989

SEPTEMBER 10

"Even the Uncertainty Principle isn't 'merely' philosophy:
it predicts real properties of electrons. Electrons jump at random from
one energy state to another state which they could never reach except that
their energy is momentarily uncertain. This 'tunneling' makes possible
the nuclear reactions that power the sun and many other processes. Physicists
have put some of these processes to practical use in microelectronics."

—DAVID CASSIDY, "HEISENBERG–QUANTUM MECHANICS,
1925–1927: WHAT GOOD IS IT?," WWW.AIP.ORG

SEPTEMBER 11

"The more I think about the physical portion of Schrödinger's theory,
the more repulsive I find it. . . . What Schrödinger writes about the visualizability
of his theory 'is probably not quite right'; in other words it's crap."

—WERNER HEISENBERG, LETTER TO WOLFGANG PAULI, 1926

SEPTEMBER 12

"Philosophers and great religious thinkers of the last century saw evidence
of God in the symmetries and harmonies around them—in the beautiful equations of
classical physics that describe such phenomena as electricity and magnetism.
I don't see the simple patterns underlying nature's complexity as evidence of God.
I believe that is God. To behold [mathematical curves], spinning to
their own music, is a wondrous, spiritual event."

—PAUL RAPP, QUOTED IN KATHLEEN MCAULIFFE'S "GET SMART: CONTROLLING CHAOS," *OMNI*, 1990

SEPTEMBER 13

"The laws of nature give a fundamental role to certain entities.
We are not really sure what they are, but at the present level of understanding they
seem to be the elementary quantum fields. They are highly simple because they are
governed by symmetries. These are not objects with which we are familiar. In fact, our
ordinary intuitive notions of space and time, causation, composition, substance,
and so on really lose their meaning on that scale. But it is just at that scale, at the level
of the quantum fields, that we are beginning to find a certain satisfying simplicity."

—STEVEN WEINBERG, "IS SCIENCE SIMPLE?" IN *THE NATURE OF THE PHYSICAL UNIVERSE*,
EDITED BY DOUGLASS HUFF AND OMER PREWETT, 1979

SEPTEMBER 14

"One might describe the mathematical quality in Nature by saying that the universe is so constituted that mathematics is a useful tool in its description. However, recent advances in physical science show that this statement of the case is too trivial. The connection between mathematics and the description of the universe goes far deeper than this, and one can get an appreciation of it only from a thorough examination of the various facts that make it up."

—PAUL DIRAC, "THE RELATION BETWEEN MATHEMATICS AND PHYSICS," *PROCEEDINGS OF THE ROYAL SOCIETY (EDINBURGH)*, 1938–1939

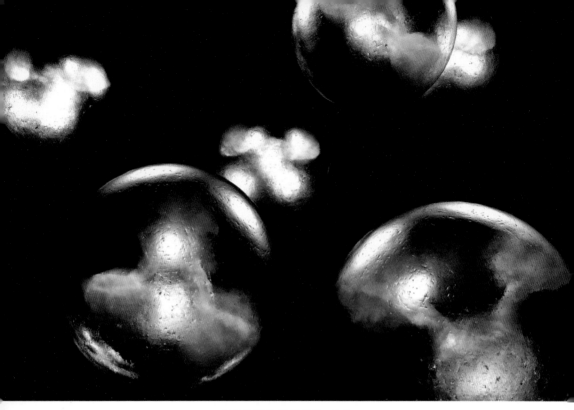

SEPTEMBER 15

BORN ON THIS DAY: Murray Gell-Mann, 1929

"For the physicist, to understand the quark is to understand the world.
The rest is just detail."

—JOSEPH SCHWARTZ, *THE CREATIVE MOMENT:
HOW SCIENCE MADE ITSELF ALIEN TO MODERN CULTURE*, 1992

SEPTEMBER 16

"How can the past and future be
when the past no longer is and the future is not yet?
As for the present, if it were always present and never moved on
to become the past, it would not be time but eternity."

—AUGUSTINE OF HIPPO, *CONFESSIONS*, C. 398

SEPTEMBER 17

"Archimedes' principle can be understood in terms of kinetic theory. . . .
When the fluid is displaced by the solid object, the molecules in the fluid will collide
with the body, exerting the same pressure as they did before the object was placed
there. For a completely submerged object . . . the molecules of the fluid will be hitting
the bottom of the object with a greater force than those hitting the top.
This is the molecular origin of the upward buoyant force."

—JAMES S. TREFIL, *THE NATURE OF SCIENCE*, 2003

itian (or self-adjoint) linear

n expressed in the represent

eralization of the Hamilton

$$i\hbar \, d\psi/dt = H\psi,$$

SEPTEMBER 18

BORN ON THIS DAY: Adrien Marie Legendre, 1752

"Most of the crackpot papers which are submitted to *The Physical Review* are rejected, not because it is impossible to understand them, but because it is possible. Those which are impossible to understand are usually published."

—FREEMAN DYSON, "INNOVATION IN PHYSICS," *SCIENTIFIC AMERICAN*, 1958

SEPTEMBER 19

"Science is about figuring out how the world works, and there are really two kinds
of science. One is where you know the rules but have to figure out how they apply in
specific situations. The other is where you try to figure out the rules themselves.
In this second category there have been revolutions—such as thermodynamics,
quantum mechanics, relativity, the genetic code—that change the whole game. . . .
The great questions lie in figuring out the rules."

—STEVEN KOONIN, "WHAT ARE THE GRAND QUESTIONS IN SCIENCE?"
IN ROBERT KUHN'S *CLOSER TO TRUTH*, 2000

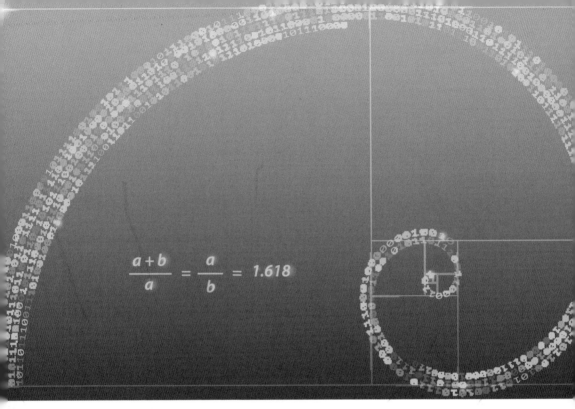

$$\frac{a+b}{a} = \frac{a}{b} = 1.618$$

SEPTEMBER 20

"The legendary Danish physicist Niels Bohr distinguished two kinds of truths. An ordinary truth is a statement whose opposite is a falsehood. A profound truth is a statement whose opposite is also a profound truth."

—FRANK WILCZEK, *THE LIGHTNESS OF BEING*, 2008

SEPTEMBER 21

"Although most physicists today place the probability of the existence of tachyons only slightly higher than the existence of unicorns, research into the properties of these hypothetical FTL [faster-than-light] particles has not been entirely fruitless."

—NICK HERBERT, *FASTER THAN LIGHT*, 1988

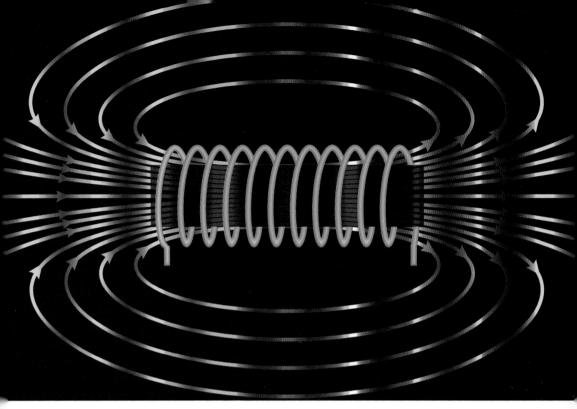

SEPTEMBER 22

BORN ON THIS DAY: Michael Faraday, 1791

"Faraday was born in the year that Mozart died. . . . Faraday's achievement is a lot less accessible than Mozart's. . . . Faraday's contributions to modern life and culture are just as great. . . . His discoveries of electromagnetic rotation and magnetic induction laid the foundations for modern electrical technology . . . and made a framework for unified field theories of electricity, magnetism, and light. . . . Faraday argued that the familiar properties of bodies reside not in matter but in forces filling all space."

—DAVID GOODING, "NEW LIGHT ON AN ELECTRIC HERO," *TIMES HIGHER EDUCATION SUPPLEMENT*, 1991

SEPTEMBER 23

"There may be further laws to discover, to do with the unification of gravity with quantum theory and with the other forces of nature. But in a certain sense, we have for the first time in history a set of laws sufficient to explain the result of every experiment that has ever been done."

—LEE SMOLIN, "NEVER SAY ALWAYS," *NEW SCIENTIST*, 2006

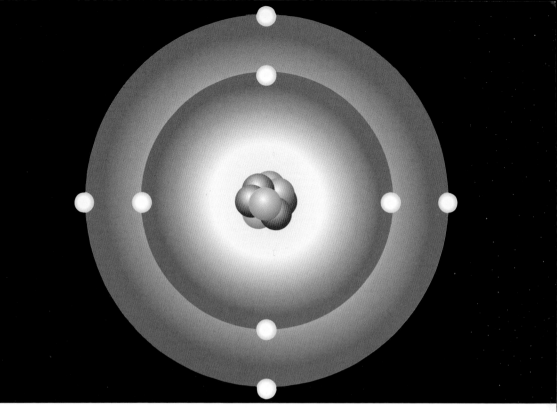

SEPTEMBER 24

"When asked whether the algorism of quantum mechanics could be considered as somehow mirroring an underlying quantum world, [Niels] Bohr would answer, 'There is no quantum world. There is only an abstract quantum physical description. It is wrong to think that the task of physics is to find out how nature is. Physics concerns what we can say about nature.'"

—AAGE PETERSEN, "THE PHILOSOPHY OF NIELS BOHR,"
BULLETIN OF THE ATOMIC SCIENTISTS, 1963

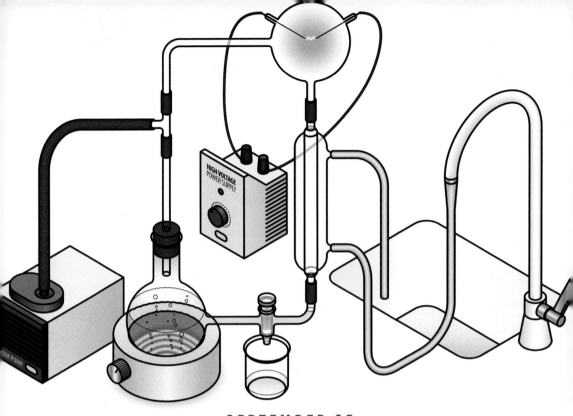

SEPTEMBER 25

"Nothing in life is certain except death, taxes and the second law of thermodynamics. All three are processes in which useful or accessible forms of some quantity, such as energy or money, are transformed into useless, inaccessible forms of the same quantity. That is not to say that these three processes don't have fringe benefits: taxes pay for roads and schools; the second law of thermodynamics drives cars, computers and metabolism; and death, at the very least, opens up tenured faculty positions."

—SETH LLOYD, "GOING INTO REVERSE," *NATURE*, 2004

SEPTEMBER 26

"Science does have some metaphysical assumptions, not the least of which is that the universe follows laws. But science leaves open the question of whether those laws were designed. That is a metaphysical question. Believing the universe or some part of it was designed or not does not help understand how it works."

—ROBERT TODD CARROLL, "INTELLIGENT DESIGN," *THE SKEPTIC'S DICTIONARY*, 2003

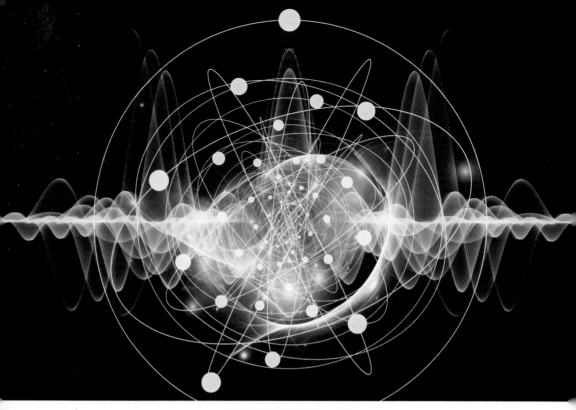

SEPTEMBER 27

"Science must be testable in principle, but that is not necessarily the same thing as testable in practice, given current technological limitations. . . . It is not uncommon for decades to go by before theories in physics are decisively confirmed. In some cases, such as the atomic theory, it has taken centuries."

—TOM SIEGFRIED, "A GREAT UNRAVELING," *NEW YORK TIMES BOOK REVIEW*, 2006

SEPTEMBER 28

"The lesson for the truth of fundamental laws is clear: fundamental laws
do not govern objects in reality; they govern only objects in models."

—NANCY CARTWRIGHT, *HOW THE LAWS OF PHYSICS LIE*, 1983

SEPTEMBER 29

BORN ON THIS DAY: Enrico Fermi, 1901

"One can argue that mathematics is a human activity deeply rooted in reality, and permanently returning to reality. From counting on one's fingers to moon-landing to Google, we are doing mathematics in order to understand, create, and handle things, . . . Mathematicians are thus more or less responsible actors of human history, like Archimedes helping to defend Syracuse (and to save a local tyrant), Alan Turing cryptanalyzing Marshal Rommel's intercepted military dispatches to Berlin, or John von Neumann suggesting high altitude detonation as an efficient tactic of bombing."

—YURI I. MANIN, "MATHEMATICAL KNOWLEDGE: INTERNAL, SOCIAL, AND CULTURAL ASPECTS," *MATHEMATICS AS METAPHOR: SELECTED ESSAYS,* 2007

SEPTEMBER 30

"In all time-travel stories where someone enters the past, the past is necessarily altered. The only way the logical contradictions created by such a premise can be resolved is by positing a universe that splits into separate branches the instant the past is entered. In other words, while time in the old branch 'gurgles on' (a phrase from Emily Dickinson) time in the new branch gurgles on in a different way toward a different future."

—MARTIN GARDNER, "CAN TIME STOP? THE PAST CHANGE?" *SCIENTIFIC AMERICAN*, 1979

OCTOBER 1

"Whence is it that the Sun and Planets gravitate towards one another,
without dense Matter between them? . . . What hinders the fix'd Stars from
falling upon one another? . . . Does it not appear from Phenomena
that there is a Being incorporeal, living, intelligent, omnipresent,
who in infinite Space . . . sees the things themselves intimately,
and thoroughly perceives them, and comprehends them wholly . . . ?"

—ISAAC NEWTON, *OPTICKS*, 1704

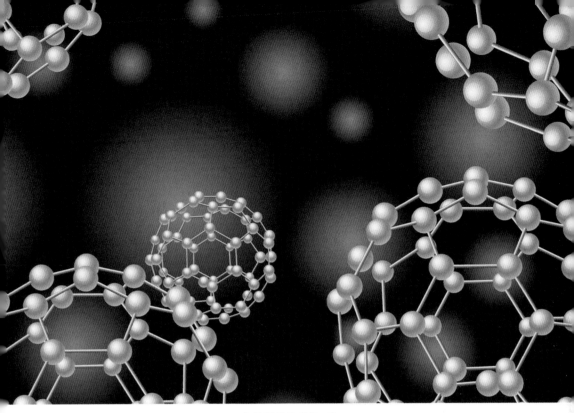

OCTOBER 2

"As a conservative, I do not agree that a division of physics into separate theories for large and small is unacceptable. I am happy with the situation in which we have lived for the last 80 years, with separate theories for the classical world of stars and planets and the quantum world of atoms and electrons."

—FREEMAN DYSON, "THE WORLD ON A STRING," *NEW YORK REVIEW OF BOOKS*, 2004

OCTOBER 3

"The tooth fairy is real, the laws of physics are real, the rules of baseball are real, and the rocks in the fields are real. But they are real in different ways. What I mean when I say that the laws of physics are real is that they are real in pretty much the same sense . . . as the rocks in the fields, and not in the same sense . . . as the rules of baseball. We did not create the laws of physics or the rocks in the field. . . . I am making an implicit assumption . . . that our statements about the laws of physics are in a one-to-one correspondence with aspects of objective reality. . . . If we ever discover intelligent creatures on some distant planet and translate their scientific works, we will find that we and they have discovered the same laws."

—STEVEN WEINBERG, "SOKAL'S HOAX," *NEW YORK REVIEW OF BOOKS*, 1996

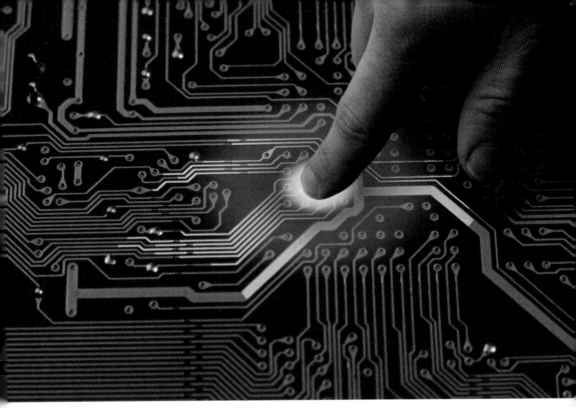

OCTOBER 4

"The operating system . . . governs the flow of information through a computer just as an eternal law of nature is thought to guide physics. But . . . there could be other kinds of architectures and operating systems that themselves evolve in time."

—LEE SMOLIN, "NEVER SAY ALWAYS," *NEW SCIENTIST*, 2006

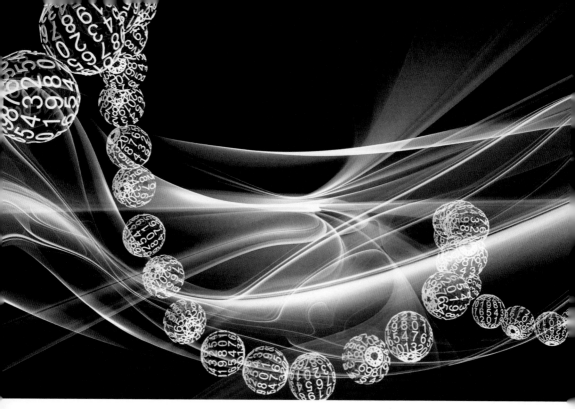

OCTOBER 5

"[God can] vary the Laws of Nature, and make Worlds of several sorts in several Parts of the Universe."

—ISAAC NEWTON, *OPTICKS*, 1718

OCTOBER 6

"Our world is filled with the statues of great generals, atop of prancing horses, leading their cheering soldiers to glorious victory. Here and there, a modest slab of marble announces that a man of science has found his final resting place. A thousand years from now we shall probably do these things differently, and the children of that happy generation shall know of the splendid courage and the almost inconceivable devotion to duty of the men who were the pioneers of that abstract knowledge, which alone has made our modern world a practical possibility."

—HENDRIK WILLEM VAN LOON, *THE STORY OF MANKIND*, 1921

OCTOBER 7

"I remember discussions with Bohr which went through many hours
till very late at night and ended almost in despair; and when
at the end of the discussion I went alone for a walk in the neighboring
park I repeated to myself again and again the question:
Can nature possibly be so absurd as it seemed to us in these atomic experiments?"

—WERNER HEISENBERG, *PHYSICS AND PHILOSOPHY: THE REVOLUTION IN MODERN SCIENCE*, 1958

OCTOBER 8

"The Higgs Boson was predicted with the same tool as the planet Neptune
and the radio wave: with mathematics. Galileo famously stated that our Universe
is a 'grand book' written in the language of mathematics. . . . I argue . . .
that our universe isn't just described by math, but that it is math in the sense that
we're all parts of a giant mathematical object, which in turn is part
of a multiverse so huge that it makes the other multiverses debated in recent
years seem puny in comparison."

—MAX TEGMARK, *OUR MATHEMATICAL UNIVERSE*, 2014

OCTOBER 9

"A good deal of my research work in physics has consisted in not setting out to solve some particular problems, but simply examining mathematical quantities of a kind that physicists use and trying to get them together in an interesting way regardless of any application that the work may have. It is simply a search for pretty mathematics. It may turn out later that the work does have an application. Then one has had good luck."

—PAUL DIRAC, "PRETTY MATHEMATICS," *INTERNATIONAL JOURNAL OF THEORETICAL PHYSICS,* 1982

OCTOBER 10

BORN ON THIS DAY: Henry Cavendish, 1731

"Mathematics may help us to measure and weigh the planets, to discover the materials of which they are composed, to extract light and warmth from the motion of water and to dominate the material universe; but even if by these means we could mount up to Mars or hold converse with the inhabitants of Jupiter or Saturn, we should be no nearer to the divine Throne, except so far as these new experiences might develop in us modesty, respect for facts, a deeper reverence for order and harmony, and a mind more open to new observations and to fresh inferences from old truths."

—EDWIN A. ABBOTT, *THE SPIRIT ON THE WATERS*, 1897

OCTOBER 11

"Profound study of nature is the most fertile source of mathematical discoveries."

—JOSEPH FOURIER, *THE ANALYTICAL THEORY OF HEAT*, 1878

OCTOBER 12

"The chess-board is the world, the pieces are the phenomena of the universe,
the rules of the game are what we call the laws of Nature. The player on the
other side is hidden from us. We know that his play is always fair, just, and patient.
But also we know, to our cost, that he never overlooks a mistake,
or makes the smallest allowance for ignorance."

—THOMAS HENRY HUXLEY, *LAY SERMONS, ADDRESSES, AND REVIEWS*, 1888

OCTOBER 13

"Mathematicians, astronomers, and physicists are often religious, even mystical; biologists much less often; economists and psychologists very seldom indeed. It is as their subject matter comes nearer to man himself that their anti-religious bias hardens."

—C. S. LEWIS, *THE GRAND MIRACLE: AND OTHER SELECTED ESSAYS ON THEOLOGY AND ETHICS FROM GOD IN THE DOCK*, 1986

OCTOBER 14

"What is the status of claims that are typically cited as 'laws of nature'—
Newton's Laws of Motion, the Law of Universal Gravitation, Snell's Laws, Ohm's Law,
the Second Law of Thermodynamics, the Law of Natural Selection? Close inspection,
I think, reveals they are neither universal nor necessary—they are not even true."

—RONALD N. GIERE, *SCIENCE WITHOUT LAWS*, 1999

OCTOBER 15

BORN ON THIS DAY: Evangelista Torricelli, 1608

"It's hardly surprising that the flow rate depends . . . on the pressure difference, pipe length, and viscosity; we've all observed that low water pressure or a long hose lengthens the time needed to fill a bucket with water and that viscous syrup pours slowly from a bottle. But the dependence of the flow rate on the fourth power of pipe diameter may come as a surprise."

—LOUIS A. BLOOMFIELD, *HOW THINGS WORK: THE PHYSICS OF EVERYDAY LIFE*, 1997

OCTOBER 16

"He watched her for a long time and she knew that he was watching her and he knew that she knew he was watching her, and he knew that she knew that he knew; in a kind of regression of images that you get when two mirrors face each other and the images go on and on and on in some kind of infinity."

—ROBERT PIRSIG, *LILA*, 1991

OCTOBER 17

"Even the briefest survey of [Isaac] Newton's life unsettles his image as the idealized prototype of a modern scientist. . . . A renowned expert on Jason's fleece, Pythagorean harmonics and Solomon's temple, his advice was also sought on the manufacture of coins and remedies for headaches. . . . Newton had no laboratory team to supervise . . . and never travelled outside eastern England. . . ."

—PATRICIA FARA, *NEWTON: THE MAKING OF GENIUS*, 2002

OCTOBER 18

"The equations [of physics] are lovely, describing how a baseball
arcs parabolically between earth and sky or how an electron jumps around
a nucleus or how a magnet pulls a pin. The ugliness is in the details.
Why does the top quark weigh roughly 40 times as much as the bottom quark . . . ?"

—GEORGE JOHNSON, "WHY IS FUNDAMENTAL PHYSICS SO MESSY?" *WIRED*, 2007

OCTOBER 19

BORN ON THIS DAY: Subrahmanyan Chandrasekhar, 1910

"Some scientists are also bothered by infinities,
which seem to crop up at embarrassing places in our theories of the universe.
A black hole . . . has an infinity at its very center."

—FRED ALAN WOLF, *PARALLEL UNIVERSES*, 1988

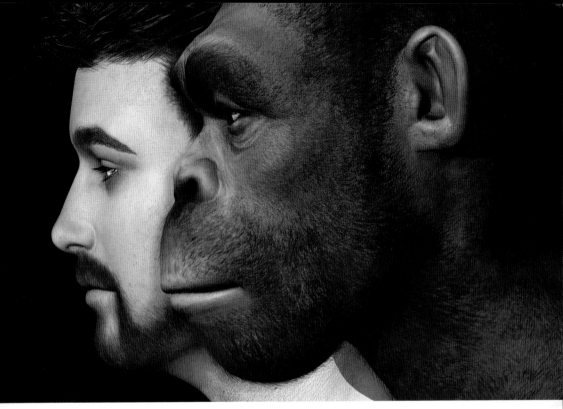

OCTOBER 20

"The fact that anything at all in the universe is comprehensible means either that we are very intelligent or that the basics of nature are very simple. Given that we are some sort of chimpanzee, carrying a mere kilogram of glop between our ears, I opt for the latter alternative."

—VINCENT ICKE, *THE FORCE OF SYMMETRY*, 1995

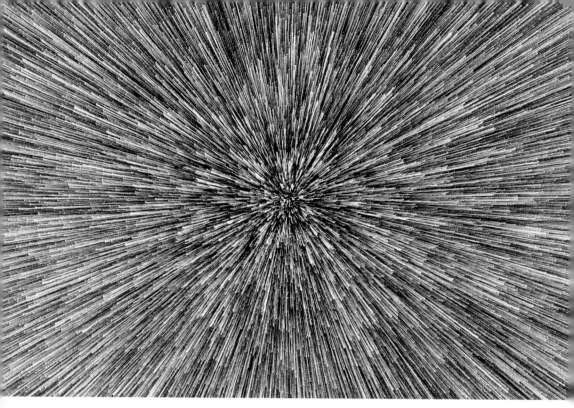

OCTOBER 21

"Arguably the most important cosmological discovery ever made
is that our Universe is expanding. [It] stands, along with the Copernican Principle . . .
and Olbers' paradox . . . that the sky is dark at night, as one of the cornerstones
of modern cosmology. It forced cosmologists to dynamic models of the Universe,
and also implies the existence of a timescale or age for the Universe. It was made
possible . . . primarily by Edwin Hubble's estimates of distances to nearby galaxies."

—JOHN P. HUCHRA, "THE HUBBLE CONSTANT," CFA.HARVARD.EDU, 2008

OCTOBER 22

"The miracle of appropriateness of the language of mathematics
for the formulation of the laws of physics is a wonderful gift which we neither
understand nor deserve. We should be grateful for it, and hope that it will remain valid
in future research and that it will extend, for better or for worse, to our pleasure even
though perhaps also to our bafflement, to wide branches of learning."

—EUGENE WIGNER, "THE UNREASONABLE EFFECTIVENESS OF MATHEMATICS IN THE NATURAL SCIENCES,"
COMMUNICATIONS ON PURE AND APPLIED MATHEMATICS, 1960

OCTOBER 23

"We all behave like Maxwell's demon. . . . We sort the mail, build sand castles, solve jigsaw puzzles, separate wheat from chaff, rearrange chess pieces . . . , and all this we do requires no great energy, as long as we can apply intelligence. . . . The original demon, discerning one molecule at a time, distinguishing fast from slow, and operating his little gateway, is sometimes described as 'superintelligent,' but compared to a real organism it is an idiot savant. Not only do living things lessen the disorder in their environments; they are in themselves, their skeletons and their flesh, vesicles and membranes, shells and carapaces, leaves and blossoms, circulatory systems and metabolic pathways—miracles of pattern and structure. It sometimes seems as if curbing entropy is our quixotic purpose in the universe."

—JAMES GLEICK, *THE INFORMATION*, 2011

OCTOBER 24

BORN ON THIS DAY: Wilhelm Weber, 1804

"Much of the history of science, like the history of religion, is a history of struggles driven by power and money. And yet this is not the whole story. Genuine saints occasionally play an important role, both in religion and in science. . . . For many scientists . . . the chief reward for being a scientist is not the power and the money but the chance of catching a glimpse of the transcendent beauty of nature."

—FREEMAN DYSON, IN THE INTRODUCTION TO *NATURE'S IMAGINATION*, EDITED BY JOHN CORNWELL, 1995

OCTOBER 25

"Imaginary time is another direction of time, one that is at right angles to ordinary, real time. We could get away from this one-dimensional, linelike behavior of time. . . . Ordinary time would be a derived concept we invent for psychological reasons. We invent ordinary time so that we can describe the universe as a succession of events in time, rather than as a static picture, like a surface map of the earth. . . . Time is just like another direction in space."

—STEPHEN HAWKING, INTERVIEW, *PLAYBOY*, 1990

OCTOBER 26

"Since Galileo, on whose discoveries much of [Isaac] Newton's own career in science would squarely rest, had died [the year of Newton's birth], a significance attaches itself to 1642. . . . Born in 1564, Galileo had lived nearly to eighty. Newton would live nearly to eighty-five. Between them, they virtually spanned the entire scientific revolution, the central core of which their combined work constituted."

—RICHARD S. WESTFALL, *NEVER AT REST: A BIOGRAPHY OF ISAAC NEWTON*, 1980

OCTOBER 27

"There is beauty in space, and it is orderly. There is no weather, and there is regularity. It is predictable. . . . Everything in space obeys the laws of physics. If you know these laws, and obey them, space will treat you kindly. And don't tell me man doesn't belong out there. Man belongs wherever he wants to go—and he'll do plenty well when he gets there."

—WERNHER VON BRAUN, "SPACE: REACH FOR THE STARS," *TIME*, 1958

OCTOBER 28

"Superstring theory has been absorbed into membrane theory, or M-theory, as they call it. There is not a scintilla of empirical evidence to support it. Although I have only a partial understanding of M-theory, it strikes me as comparable to Ptolemy's epicycles. It's getting more and more baroque."

—MARTIN GARDNER IN "INTERVIEW WITH MARTIN GARDNER," *NOTICES OF THE AMERICAN MATHEMATICAL SOCIETY*, 2005

OCTOBER 29

"Armies of thinkers have been defeated by the enigma of why most fundamental laws of nature can be written down so conveniently as equations. Why is it that so many laws can be expressed as an absolute imperative, that two apparently unrelated quantities (the equation's right and left sides) are exactly equal? Nor is it clear why fundamental laws exist at all."

—GRAHAM FARMELO, FOREWORD TO *IT MUST BE BEAUTIFUL*, 2003

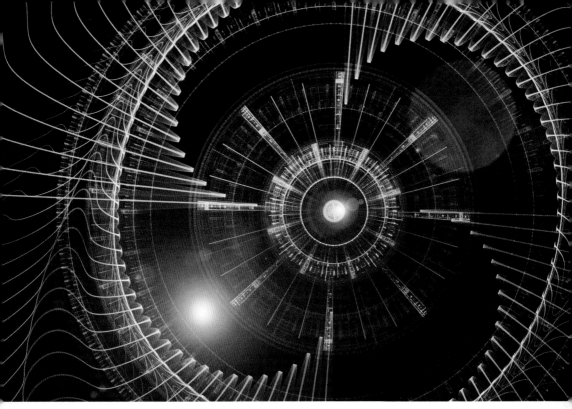

OCTOBER 30

"Every atom being self-existent, had the power in the beginning to adopt what laws of motion it pleased; so they all, by some mysterious universal suffrage conveyed through the infinity of space . . . mutually agreed on the law and intensity of gravity, and have steadily kept to their agreement ever since. If this proposition looks absurd, atheists can blame no one but themselves, for the doctrine of inherent forces cannot be translated in plain English in any other way."

—HENRY AUGUSTUS MOTT, *THE LAWS OF NATURE*
AND MAN'S POWER TO MAKE THEM SUBSERVIENT TO HIS WISHES, 1882

OCTOBER 31

"The universe might actually be able to fine-tune itself.
If you assume the laws of physics do not reside outside the physical universe,
but rather are part of it, they can only be as precise as can be calculated
from the total information content of the universe. The universe's information
content is limited by its size, so just after the big bang, while the universe
was still infinitesimally small, there may have been wiggle room,
or imprecision, in the laws of nature."

—PATRICK BARRY, "WHAT'S DONE IS DONE . . . OR IS IT?," *NEW SCIENTIST*, 2006

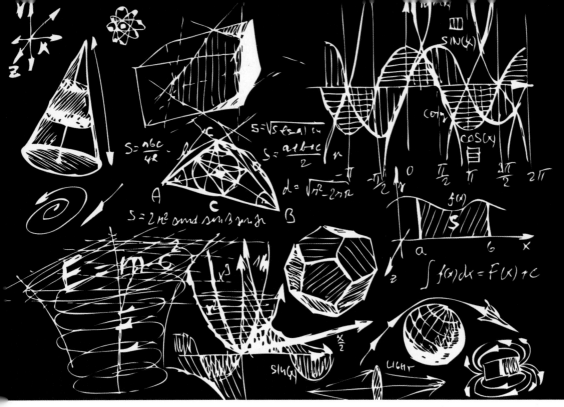

NOVEMBER 1

"If our abstractions of nature are mathematical,
in what sense can we be said to understand the universe?
For example, in what sense does Newton's Law explain why things move?"

—LAWRENCE M. KRAUSS, *FEAR OF PHYSICS*, 1993

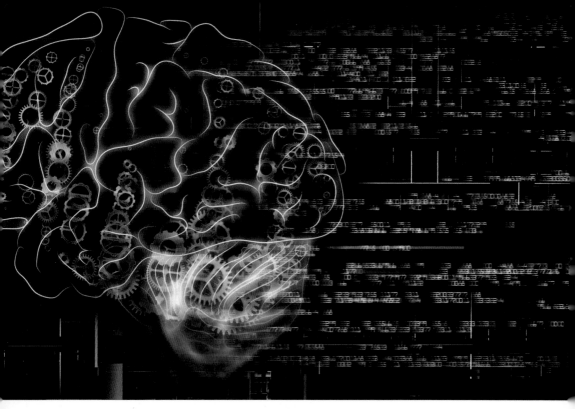

NOVEMBER 2

"We see reality according to our thought. Therefore thought is constantly participating both in giving shape and form and figuration to ourselves, and to the whole of reality. Now, thought doesn't know this. Thought is thinking that it isn't doing anything. I think this is really where the difficulty is. We have got to see that thought is part of this reality and that we are not merely thinking about it, but that we are thinking it. Do you see the difference?"

—DAVID BOHM, *ON CREATIVITY*, EDITED BY LEE NICHOL, 1996

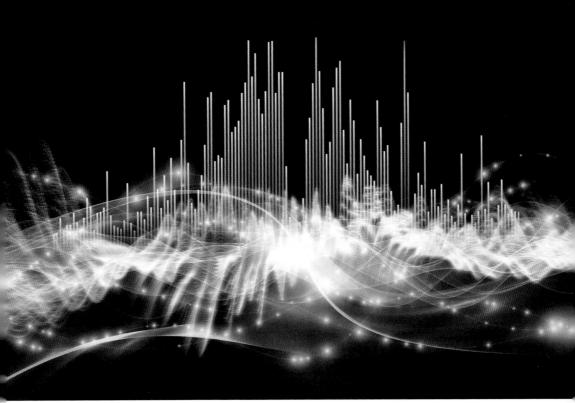

NOVEMBER 3

"Physics is really nothing more than a search for ultimate simplicity,
but so far all we have is a kind of elegant messiness."

—BILL BRYSON, *A SHORT HISTORY OF NEARLY EVERYTHING*, 2004

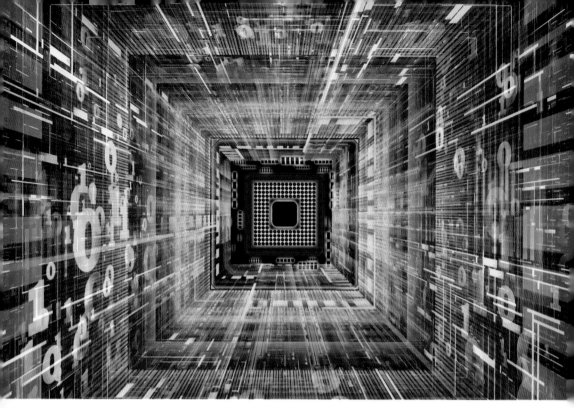

NOVEMBER 4

"There is a philosophy that says that if something is unobservable—
unobservable in principle—it is not part of science. If there is no way to falsify
or confirm a hypothesis, it belongs to the realm of metaphysical speculation,
together with astrology and spiritualism. By that standard, most of the universe
has no scientific reality—it's just a figment of our imaginations."

—LEONARD SUSSKIND, *THE BLACK HOLE WAR*, 2008

NOVEMBER 5

"Every theoretical physicist who is any good knows six or seven
different theoretical representations for exactly the same physics.
He knows that they are all equivalent, and that nobody is ever going to be
able to decide which one is right at that level, but he keeps them in his head,
hoping that they will give him different ideas for guessing."

—RICHARD FEYNMAN, *THE CHARACTER OF PHYSICAL LAW*, 1965

NOVEMBER 6

"Descartes and Newton both thought of laws of nature [as being imposed by God]. . . . According to this traditional view, material things are bound to act according to the laws of nature, because such things are themselves powerless. Only spiritual beings, such as God, human minds, or angels, were thought to be capable of acting on their own account. . . . In the eighteenth century, secularized versions of the divine command theory were developed. Instead of thinking of God as the source of all power and order, some natural philosophers of the period began to speak of the 'forces of nature' as the source of nature's activity."

—BRIAN DAVID ELLIS, *THE PHILOSOPHY OF NATURE: A GUIDE TO THE NEW ESSENTIALISM*, 2002

NOVEMBER 7

BORN ON THIS DAY: Marie Skłodowska-Curie, 1867; Lise Meitner, 1878

"As a man who has devoted his whole life to the most clear-headed science,
to the study of matter, I can tell you as a result of my research about atoms this much:
There is no matter as such. All matter originates and exists only by virtue of
a force which brings the particle of an atom to vibration and holds this most minute
solar system of the atom together. We must assume behind this force the existence
of a conscious and intelligent mind. This mind is the matrix of all matter."

—MAX PLANCK, "DAS WESEN DER MATERIE (THE NATURE OF MATTER)," LECTURE, FLORENCE, ITALY, 1944

NOVEMBER 8

"I call our world Flatland, not because we call it so, but to make its nature
clearer to you, my happy readers, who are privileged to live in Space.
Imagine a vast sheet of paper on which straight Lines, Triangles, Squares, Pentagons,
Hexagons, and other figures, instead of remaining fixed in their places,
move freely about, on or in the surface, but without the power of rising above or
sinking below it, very much like shadows . . . and you will then have a pretty correct
notion of my country and countrymen. Alas! a few years ago, I should have said
'my universe,' but now my mind has been opened to higher views of things."

—EDWIN A. ABBOTT, *FLATLAND: A ROMANCE OF MANY DIMENSIONS*, 1884

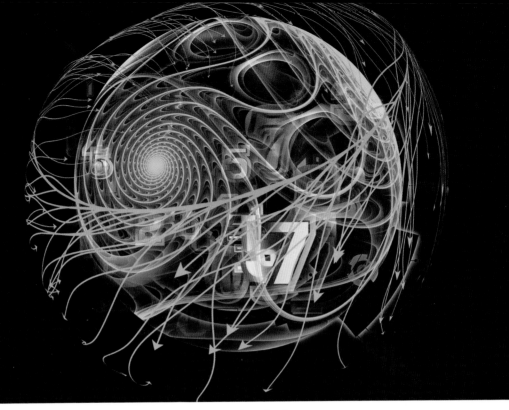

NOVEMBER 9

"Adapting from the earlier book *Gravitation*, I wrote, 'Spacetime tells matter how to move; matter tells spacetime how to curve.' In other words, a bit of matter (or mass, or energy) moves in accordance with the dictates of the curved spacetime where it is located. . . . At the same time, that bit of mass or energy is itself contributing to the curvature of spacetime everywhere."

—JOHN ARCHIBALD WHEELER WITH KENNETH WILLIAM FORD,
GEONS, BLACK HOLES, AND QUANTUM FOAM, 2000

NOVEMBER 10

"At once philosophy and genial fantasy, practical physics and terrifying weapon,
$E = mc^2$ has become metonymic of technical knowledge writ large.
Our ambitions for science, our dreams of understanding and our nightmares
of destruction find themselves packed into a few scribbles of the pen."

—PETER GALISON, "THE SEXTANT EQUATION," IN GRAHAM FARMELO'S *IT MUST BE BEAUTIFUL*, 2003

NOVEMBER 11

"It took less than an hour to make the atoms, a few hundred million years to make the stars and planets, but five billion years to make man!"

—GEORGE GAMOW, *THE CREATION OF THE UNIVERSE*, 1952

NOVEMBER 12

"If we wish to understand the nature of the Universe,
we have an inner hidden advantage: we are ourselves little portions
of the universe and so carry the answer within us."

—JACQUES BOIVIN, *THE SINGLE HEART FIELD THEORY*, 1978

NOVEMBER 13

"If you're willing to answer yes to a God outside of nature,
then there's nothing inconsistent with God on rare occasions choosing to
invade the natural world in a way that appears miraculous.
If God made the natural laws, why could he not violate them
when it was a particularly significant moment for him to do so?"

—FRANCIS COLLINS, "GOD VS. SCIENCE" (INTERVIEW), *TIME*, 2006

NOVEMBER 14

"Laws of nature are those fundamental laws of physics which hold everywhere within the universe. . . . There are only a few of them, [e.g.,] $F(x,t) = m(x) \cdot d^2 s(x,t)/dt^2$ . . . Another kind of law of nature are *special force laws*, e.g., the classical laws for gravitational force or electric force. . . . Laws of nature are *strictly* true . . . —but at the cost of not per se being *applicable* to *real* systems, because they do not specify *which* forces are active. *System laws*, in contrast, . . . refer to particular systems of a certain kind in a certain time interval Δt. They contain or rely on a *specification* of all forces . . . Examples of system laws in classical physics are Kepler's laws of elliptic planetary orbits, Galileo's law of falling bodies, the classical wave equations, etc."

—GERHARD SCHURZ, "NORMIC LAWS, NON-MONOTONIC REASONING, AND UNITY OF SCIENCE," IN *LOGIC, EPISTEMOLOGY, AND THE UNITY OF SCIENCE*, EDITED BY S. RAHMAN ET AL., 2004

NOVEMBER 15

"It seems we all face a fundamental paradox in that it's impossible to think about the universe except in terms of its relation to humans. You can't make sense of language, or even scientific laws or mathematics, without the concept of an observer, and yet at the same time we know perfectly well that humans are a very late addition: the universe was here long before us and will be here long after us."

—MICHAEL FRAYN, "ALL THE WORLD'S A STAGE," INTERVIEW WITH LIZ ELSE, *NEW SCIENTIST*, 2006

NOVEMBER 16

"One moonlit night we walked all over the Hainberg Mountain [near Göttingen], and [Heisenberg] was completely enthralled by the visions he had, trying to explain his newest discovery to me. He talked about the miracle of symmetry as the original archetype of creation, about harmony, about the beauty of simplicity, and its inner truth. It was a high point of our lives."

—ELISABETH HEISENBERG, *INNER EXILE*, 1984

NOVEMBER 17

"Imagine that the world is something like a great chess game being played by the gods, and we are observers of the game. . . . If we watch long enough, we may eventually catch on to a few of the rules. . . . However, we might not be able to understand why a particular move is made in the game, merely because it is too complicated and our minds are limited. . . . We must limit ourselves to the more basic question of the rules of the game. If we know the rules, we consider that we 'understand' the world."

—RICHARD FEYNMAN, *THE FEYNMAN LECTURES ON PHYSICS*, 1964

NOVEMBER 18

"While the equations represent the discernment of eternal and universal truths, however, the manner in which they are written is strictly, provincially human. That is what makes them so much like poems, wonderfully artful attempts to make infinite realities comprehensible to finite beings."

—MICHAEL GUILLEN, *FIVE EQUATIONS THAT CHANGED THE WORLD*, 1995

NOVEMBER 19

"You may not feel outstandingly robust, but if you are an average-sized adult you will contain within your modest frame no less than 7×10^{18} joules of potential energy—enough to explode with the force of thirty very large hydrogen bombs, assuming you knew how to liberate it and really wished to make a point."

—BILL BRYSON, *A SHORT HISTORY OF NEARLY EVERYTHING*, 2004

NOVEMBER 20

BORN ON THIS DAY: Edwin Hubble, 1889

"[Hubble] opened the world of galaxies for science when he proved
that the nebulae outside the Milky Way are gigantic stellar systems [similar to]
the galaxy which includes our Sun and its planets. However, the most important
discovery was that of the red-shift in spectra of galaxies. . . . The Universe was
smaller in the past. . . . The explosive origin of the universe determined
its subsequent evolution [that] gave rise to the human race. . . . This is why
astronomers rank Edwin Hubble with Copernicus and Galileo Galilei."

—ALEXANDER S. SHAROV AND IGOR D. NOVIKOV,
EDWIN HUBBLE, THE DISCOVERER OF THE BIG BANG UNIVERSE, 2005

NOVEMBER 21

"The essential fact is simply that all the pictures which science now draws of nature . . . are mathematical pictures. . . . It can hardly be disputed that nature and our conscious mathematical minds work according to the same laws."

—JAMES HOPWOOD JEANS, *THE MYSTERIOUS UNIVERSE*, 1930

POSTA ROMANA

55 B

400 ANI DE LA NASTERE
1971

KEPLER

VLASTO

NOVEMBER 22

"Although the principal idea of the *Mysterium cosmographicum*
[an astronomy book by Johannes Kepler] was erroneous,
Kepler established himself as the first . . . scientist to demand physical
explanations for celestial phenomena. Seldom in history has so wrong
a book been so seminal in directing the future course of science."

—OWEN GINGERICH, "KEPLER," *DICTIONARY OF SCIENTIFIC BIOGRAPHY*, 1981

NOVEMBER 23

"It is now generally accepted that the universe evolves according to well-defined laws. These laws may have been ordained by God, but it seems that He does not intervene in the universe to break the laws."

—STEPHEN HAWKING, *BLACK HOLES AND BABY UNIVERSES AND OTHER ESSAYS*, 1993

NOVEMBER 24

"The poetry of science is in some sense embodied in its great equations and . . . these equations can also be peeled. But their layers represent their attributes and consequences, not their meanings. . . . It is perfectly possible to imagine a universe in which mathematical equations have nothing to do with the workings of nature. Yet the marvellous thing is that they do."

—GRAHAM FARMELO, FOREWORD TO *IT MUST BE BEAUTIFUL*, 2003

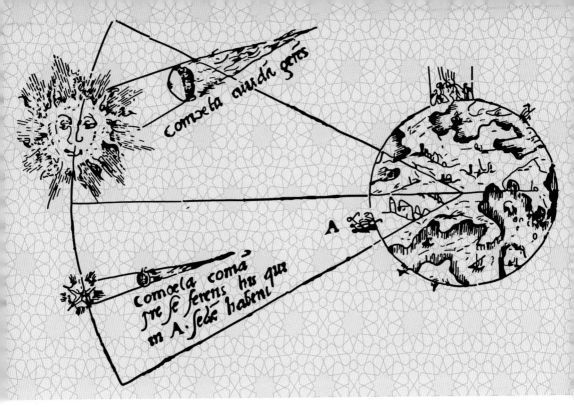

NOVEMBER 25

"None of the laws of physics known today
(with the possible exception of the general principles of quantum mechanics)
are exactly and universally valid. Nevertheless, many of them have settled
down to a final form, valid in certain known circumstances."

—STEVEN WEINBERG, "SOKAL'S HOAX," *NEW YORK REVIEW OF BOOKS*, 1996

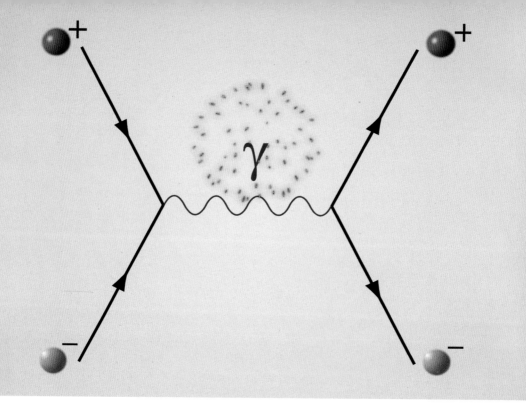

NOVEMBER 26

"If you see an antimatter version of yourself rushing toward you,
think twice before embracing."

—J. RICHARD GOTT III, *TIME TRAVEL IN EINSTEIN'S UNIVERSE*, 2001

NOVEMBER 27

"Newton was not the first of the age of reason. He was the last of the magicians, the last of the Babylonians and Sumerians, the last great mind which looked out on the visible and intellectual world with the same eyes as those who began to build our intellectual inheritance rather less than 10,000 years ago. . . . [Newton saw] the whole universe and all that is in it as a riddle, as a secret which could be read by applying pure thought to certain evidence, certain mystic clues which God had laid about the world to allow a sort of philosopher's treasure hunt. . . . He regarded the universe as a cryptogram set by the Almighty. . . ."

–JOHN MAYNARD KEYNES, "ESSAYS IN BIOGRAPHY: NEWTON, THE MAN,"
THE COLLECTED WRITINGS OF JOHN MAYNARD KEYNES, 1972

NOVEMBER 28

"Before creation, God did just pure mathematics.
Then He thought it would be a pleasant change to do some applied."

—JOHN EDENSOR LITTLEWOOD, *A MATHEMATICIAN'S MISCELLANY*, 1953

NOVEMBER 29

BORN ON THIS DAY: Christian Doppler, 1803

"And from my pillow, looking forth by light
Of moon or favouring stars, I could behold
The antechapel where the statue stood
Of Newton with his prism and silent face,
The marble index of a mind for ever
Voyaging through strange seas of Thought, alone."

—WILLIAM WORDSWORTH, *THE PRELUDE*, BOOK III, 1805

NOVEMBER 30

"According to the general theory of relativity, the laws of motion
can be expressed in any inertial or accelerated frame. Thus the choice
between a heliocentric model and an earth-centric one is not a matter
of right or wrong but one of convention and convenience. What is a model
assumption and what is convention is not always clear."

—BYRON K. JENNINGS, "ON THE NATURE OF SCIENCE," *PHYSICS IN CANADA*, 2007

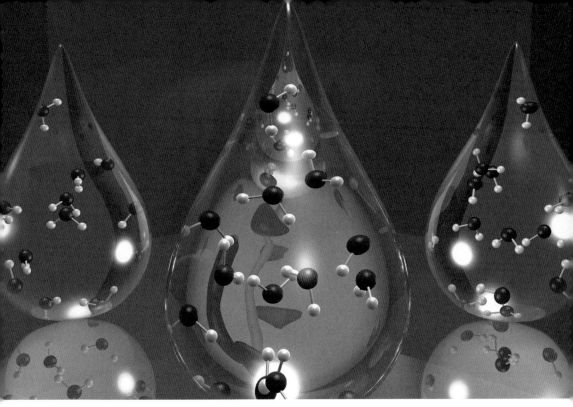

DECEMBER 1

"You find that many laws of nature are themselves connected to others by still deeper laws, that those deeper laws have even deeper connections, and so on. Eventually, at the very center of the web, you find a relatively small number of laws that cement the whole framework together. . . . These are sometimes referred to as 'laws of nature'. . . . To paraphrase Animal Farm, all laws of nature are equal, but some laws are more equal than others. . . . Of course, there is no universal agreement among scientists as to what the overarching principles of our craft are exactly, but you would be hard pressed to find a scientist who doesn't agree that they exist."

—JAMES S. TREFIL, *THE NATURE OF SCIENCE*, 2003

DECEMBER 2

"Physicists have come to realize that mathematics, when used with sufficient care, is a proven pathway to truth."

—BRIAN GREENE, *THE FABRIC OF THE COSMOS*, 2004

DECEMBER 3

"Why should the universe be constructed in such a way
that atoms acquire the ability to be curious about themselves?"

—MARCUS CHOWN, *THE MAGIC FURNACE*, 1999

DECEMBER 4

"They came again, so many of them but this time I only smiled and I didn't open my eyes. You can come, you aren't going to make me jump and wake up. No, you can come, even if there are so many of you there are no numbers for you. You come from the place where there are no numbers."

—ANNE RICE, *CHRIST THE LORD*, 2008

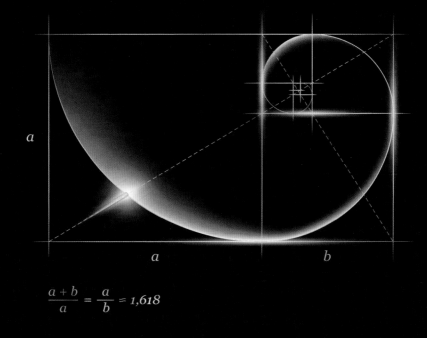

$$\frac{a+b}{a} = \frac{a}{b} \approx 1{,}618$$

DECEMBER 5

BORN ON THIS DAY: Werner Heisenberg, 1901

"Heisenberg once made the following remark to Einstein:
'If nature leads us to mathematical forms of great simplicity and beauty . . .
that no one has previously encountered, we cannot help thinking that they are "true,"
that they reveal a genuine feature of nature.'"

—PAUL DAVIES, *SUPERFORCE*, 1985

DECEMBER 6

"What then is time? If no one asks me, I know what it is.
If I wish to explain it to him who asks, I do not know."

—AUGUSTINE OF HIPPO, *CONFESSIONS*, C. 398

DECEMBER 7

"Let us now discuss the extent of the mathematical quality in Nature.
According to the mechanistic scheme of physics or to its relativistic modification,
one needs for the complete description of the universe not merely a complete system
of equations of motion, but also a complete set of initial conditions, and it is only
to the former of these that mathematical theories apply. The latter are considered to be
not amenable to theoretical treatment and to be determinable only from observation."

—PAUL DIRAC, "THE RELATION BETWEEN MATHEMATICS AND PHYSICS,"
PROCEEDINGS OF THE ROYAL SOCIETY (EDINBURGH), 1938-1939

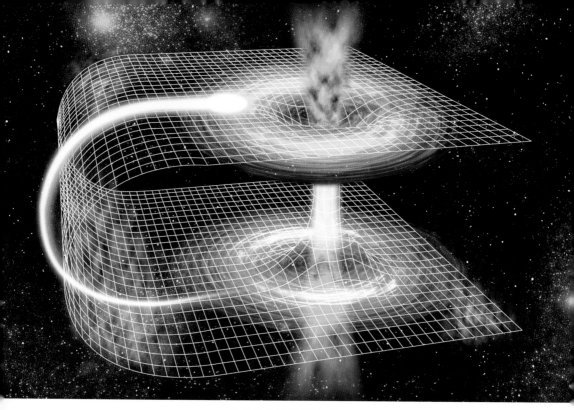

DECEMBER 8

"Einstein's theory links space and time into an inseparable unity.
As a result, any wormhole that connects two distant points in space
might also connect two distant points in time. In other words,
Einstein's theory allows for the possibility of time travel."

—MICHIO KAKU, *PARALLEL WORLDS*, 2006

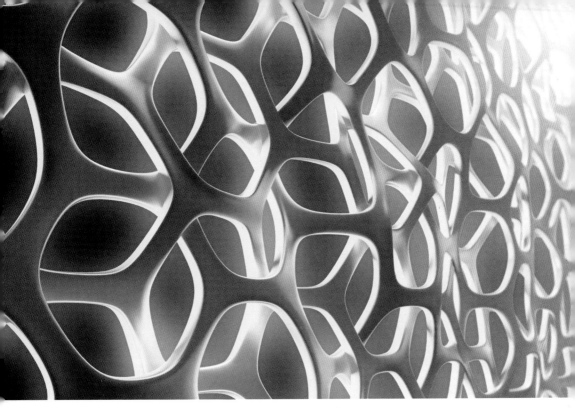

DECEMBER 9

"All science is either physics or stamp collecting."

—ERNEST RUTHERFORD, QUOTED IN J. B. BIRKS'S *RUTHERFORD AT MANCHESTER*, 1962

DECEMBER 10

"How wonderful that we have met with a paradox.
Now we have some hope of making progress."

—NIELS BOHR, AS QUOTED IN RUTH MOORE'S *NIELS BOHR: THE MAN, HIS SCIENCE,
& THE WORLD THEY CHANGED*, 1966

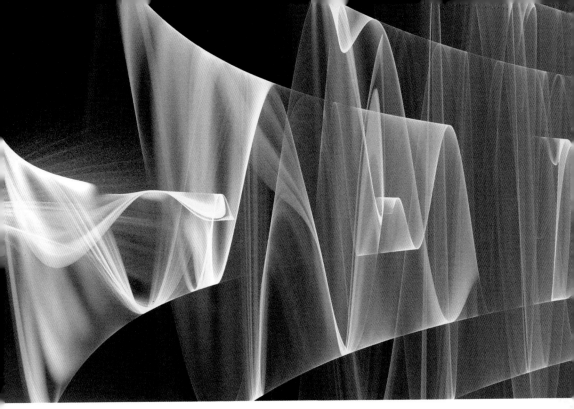

DECEMBER 11

"Heat can evidently be a cause of motion only by virtue
of the changes of volume or of form which it produces in bodies.
These changes are not caused by uniform temperature,
but rather by alternations of heat and cold."

—NICOLAS LÉONARD SADI CARNOT, *REFLECTIONS ON THE MOTIVE POWER OF HEAT
AND ON MACHINES FITTED TO DEVELOP POWER*, 1824

DECEMBER 12

"Energy is liberated matter; matter is energy waiting to happen."

—BILL BRYSON, *A SHORT HISTORY OF NEARLY EVERYTHING*, 2004

DECEMBER 13

"When we say two bodies 'touch,' what we mean (without knowing it)
is that both electromagnetic fields are interacting to avoid physical interpenetration
and . . . that happens well before subatomic particles touch!"

—FELIX ALBA-JUEZ, *GALLOPING WITH LIGHT*, 2010

DECEMBER 14

"After the Christian had come to believe that something was true simply because God had revealed it, the believer could analyze that truth philosophically in an attempt to determine how it could be true. Further, reason was seen as playing an important role in convincing atheists of the truth of Christianity by showing that the mysteries of Christianity are rationally possible . . ."

—JAN W. WOJCIK, *ROBERT BOYLE AND THE LIMITS OF REASON*, 2002

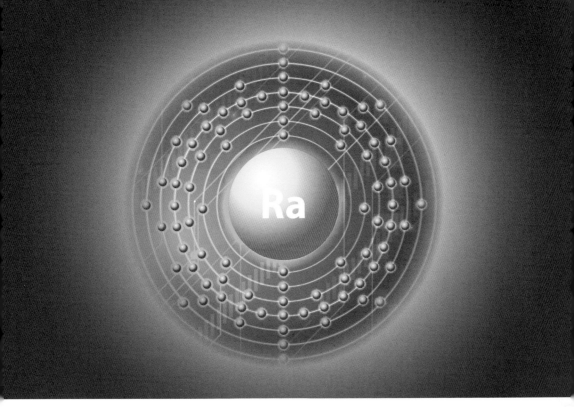

DECEMBER 15

BORN ON THIS DAY: Henri Becquerel, 1852

"Radium could become very dangerous in criminal hands, and here the question can be raised whether mankind benefits from knowing the secrets of Nature. . . ."

—PIERRE CURIE, "RADIOACTIVE SUBSTANCES, ESPECIALLY RADIUM"
(NOBEL PRIZE LECTURE), STOCKHOLM, 1905

DECEMBER 16

"Physics tries to discover the pattern of events which controls
the phenomena we observe. But we can never know what this pattern means
or how it originates; and even if some superior intelligence were to tell us,
we should find the explanation unintelligible."

—JAMES HOPWOOD JEANS, *PHYSICS AND PHILOSOPHY*, 1942

DECEMBER 17

"We have a closed circle of consistency here: the laws of physics produce complex systems, and these complex systems lead to consciousness, which then produces mathematics, which can then encode in a succinct and inspiring way the very underlying laws of physics that gave rise to it."

—PAUL DAVIES, *ARE WE ALONE?*, 1995

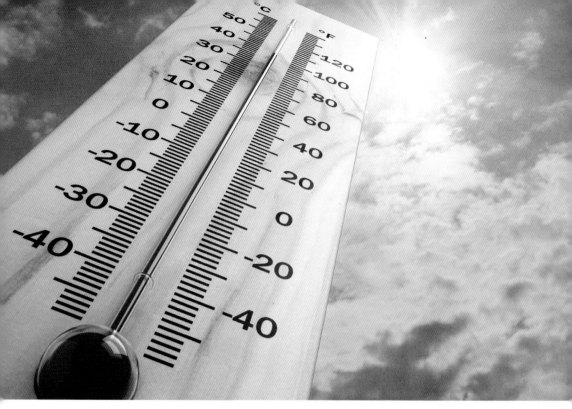

DECEMBER 18

BORN ON THIS DAY: Joseph John "J. J." Thomson, 1856

"The history of thermodynamics is a story of people and concepts.
The cast of characters is large. At least ten scientists played major roles in creating
thermodynamics, and their work spanned more than a century. The list of concepts,
on the other hand, is surprisingly small; there are just three leading concepts in
thermodynamics: energy, entropy, and absolute temperature."

— WILLIAM H. CROPPER, *GREAT PHYSICISTS*, 2004

DECEMBER 19

"Of all obstacles to a thoroughly penetrating account of existence, none looms up more dismayingly than 'time.' Explain time? Not without explaining existence. Explain existence? Not without explaining time. To uncover the deep and hidden connection between time and existence . . . is a task for the future."

—JOHN ARCHIBALD WHEELER,
"HERMANN WEYL AND THE UNITY OF KNOWLEDGE," *AMERICAN SCIENTIST*, 1986

DECEMBER 20

"But to believe a law is useful and reliable is not the same thing as to believe it is eternally true. We could just as easily believe there is nothing but an infinite succession of approximate laws. Or that laws are generalisations about nature that are not unchanging, but change so slowly that until now we have imagined them as eternal."

—LEE SMOLIN, "NEVER SAY ALWAYS," *NEW SCIENTIST*, 2006

DECEMBER 21

"Superstring theory turns out to be more complex than the universe it is supposed to simplify. Research suggests there may be 10^{500} universes . . . each ruled by different laws. The truths that Newton, Einstein, and dozens of lesser lights have uncovered would be no more fundamental than the municipal code of Nairobi. . . . Physicists would just be geographers of some accidental terrain."

—GEORGE JOHNSON, "WHY IS FUNDAMENTAL PHYSICS SO MESSY?," *WIRED*, 2007

DECEMBER 22

"When we look back at the scientific revolution from our vantage point of three centuries and attempt to understand the momentous transformation of Western thought by isolating its central characteristic, the ever greater role of mathematics and of quantitative modes of thought insistently catches our eye—what Alexandre Koyré dubbed the geometrization of nature. Initiated in the sixteenth and seventeenth centuries, the geometrization of nature has proceeded with gathering momentum ever since. To be a scientist today is to understand and to do mathematics; such is perhaps our most distinctive legacy from the scientific revolution."

—RICHARD S. WESTFALL, "NEWTON'S SCIENTIFIC PERSONALITY," *JOURNAL OF THE HISTORY OF IDEAS*, 1987

DECEMBER 23

"If [a coin] comes down heads, that means that the possibility
of its coming down tails has collapsed. Until that moment the two possibilities
were equal. But on another world, it does come down tails.
And when that happens, the two worlds split apart."

—PHILLIP PULLMAN, *THE GOLDEN COMPASS*, 1995

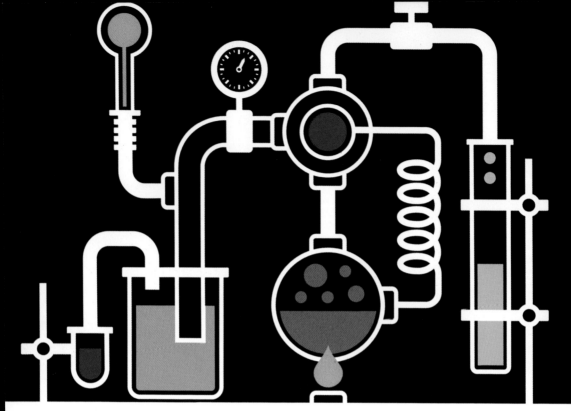

DECEMBER 24

BORN ON THIS DAY: James Prescott Joule, 1818

"In the vestibule of the Manchester Town Hall are placed two life-sized marble statues facing each other. . . . Thus honour is done to Manchester's two greatest sons—to Dalton, the founder of modern Chemistry and of the Atomic Theory, and the discoverer of the laws of chemical-combining proportions; to Joule, the founder of modern Physics and the discoverer of the law of the Conservation of Energy. The one gave to the world the final and satisfactory proof . . . that in every kind of chemical change no loss of matter occurs; the other proved that in all the varied modes of physical change no loss of energy takes place."

—HENRY E. ROSCOE, *JOHN DALTON AND THE RISE OF MODERN CHEMISTRY*, 1895

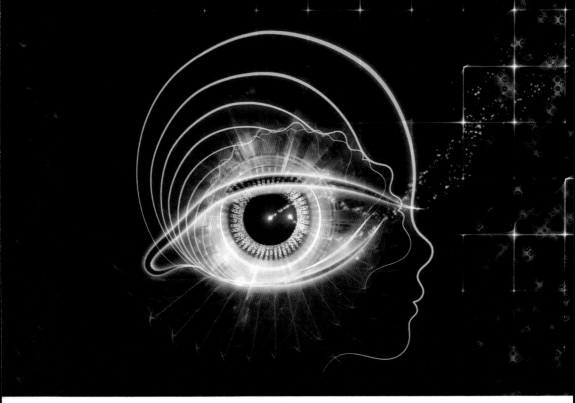

DECEMBER 25

BORN ON THIS DAY: Isaac Newton, 1642

"Nature, and Nature's laws lay hid in night. God said:
'Let Newton be!' and all was light."

—ALEXANDER POPE, "EPITAPH INTENDED FOR SIR ISAAC NEWTON,"
IN *THE COMPLETE POETICAL WORKS OF POPE*, 1931

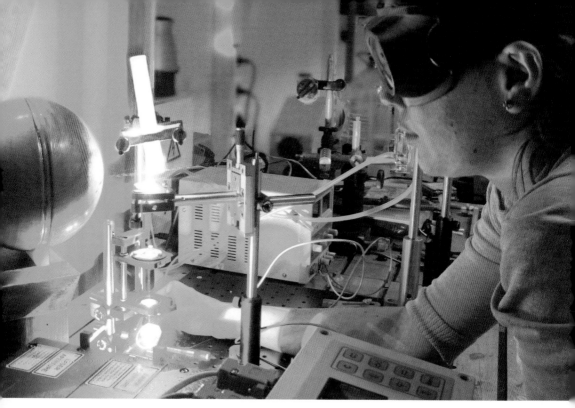

DECEMBER 26

"Mathematicians are only dealing with the structure of reasoning,
and they do not really care what they are talking about.
They do not even need to know what they are talking about . . .
But the physicist has meaning to all his phrases. . . . In physics, you have to
have an understanding of the connection of words with the real world."

—RICHARD FEYNMAN, *THE CHARACTER OF PHYSICAL LAW*, 1965

DECEMBER 27

BORN ON THIS DAY: Johannes Kepler, 1571

"Although Kepler is remembered today chiefly for his three laws of planetary motion, these were but three elements in his much broader search for cosmic harmonies. . . . He left [astronomy] with a unified and physically motivated heliocentric system nearly 100 times more accurate."

—OWEN GINGERICH, "KEPLER," *DICTIONARY OF SCIENTIFIC BIOGRAPHY*, 1981

DECEMBER 28

"The possibilities of space travel beckon us every time we gaze up at the stars, yet we seem to be permanent captives in the present. The question that motivates not only dramatic license but a surprising amount of modern theoretical physics research can be simply put: Are we or are we not prisoners on a cosmic temporal freight train that cannot jump the tracks?"

—LAWRENCE M. KRAUSS, *THE PHYSICS OF STAR TREK*, 2007

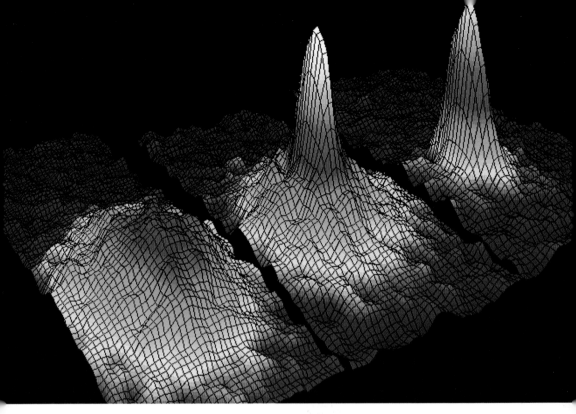

DECEMBER 29

"It is often stated that of all the theories proposed in this century,
the silliest is quantum theory. Some say that the only thing that quantum theory
has going for it, in fact, is that it is unquestionably correct."

—MICHIO KAKU, *HYPERSPACE*, 1995

DECEMBER 30

"Each of us is aware he's a material being, subject to the laws of physiology and physics, . . . The eternal belief of lovers and poets in the power of love, which is more enduring than death, the *finis vitae sed non amoris* that has pursued us through the centuries, is a lie. But this lie is not ridiculous, it's simply futile. To be a clock, on the other hand, measuring the passage of time, one that is smashed and rebuilt over and again, one in whose mechanism despair and love are set in motion by the watchmaker along with the first movements of the cogs. To know one is a repeater of suffering felt ever more deeply as it becomes increasingly comical through multiple repetitions. To replay human existence—fine, but to replay it in the way a drunk replays a corny tune, pushing coins over and over into the jukebox?"

—STANISŁAW LEM, *SOLARIS*, 1961

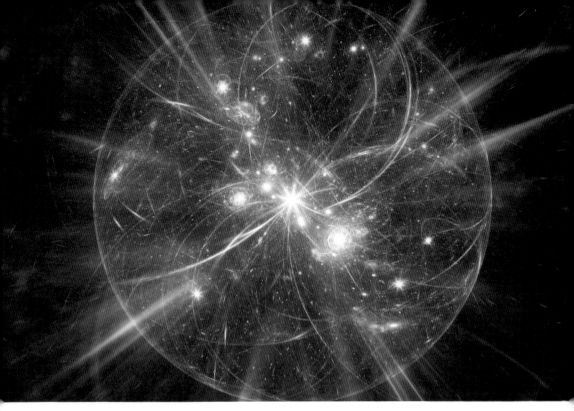

DECEMBER 31

"How did the universe know, at that moment of beginning,
what laws to follow?"

—LEE SMOLIN, "NEVER SAY ALWAYS," *NEW SCIENTIST*, 2006

BIOGRAPHICAL NOTES

Think of the following micro-biographies for the physicists featured in "Born on This Day" entries as a kind of curiosity file, highlighting some of the advanced fields in which the physicist worked, with an occasional equation that is in some way related to his or her research or contribution. Due to constraints of space, equations are generally not explained but are meant to serve as a launchpad for further exploration in other books and on the Web. For several physicists, I have included the title of one or two of their many publications.

I apologize for the very short and choppy approach, without many descriptions. Again, this is done purely in the interest of brevity. Nevertheless, I think many of the facts and formulas may pique your curiosity and encourage you to further explore a topic or a scientist's life.

In closing, I thank Dennis Gordon and Jessica Giordano for useful feedback during the preparation of this book.

AMPÈRE, ANDRÉ-MARIE (1775–1836), FRANCE

Ampère was one of the primary founders of the study of classical electromagnetism. He is famous for his law showing that the magnetic circulation in free space, along concentric paths around a straight wire carrying a current, is proportional to the total current through the surface bounding the path over which the circulation is computed. Ampère's law is expressed in many forms, perhaps most famously with the integral calculus equation:

$$\oint_s \mathbf{B} \cdot d\mathbf{s} = \mu_0 I_{\text{enc}}$$

where $\mathbf{B}$ is the magnetic field, and I is the current.

BARDEEN, JOHN (1908–1991), UNITED STATES

Bardeen contributed to the invention of the transistor and to a fundamental theory of superconductivity.

BECQUEREL, HENRI (1852–1908), FRANCE

Becquerel was one of the discoverers of radioactivity, along with Marie Skłodowska-Curie and Pierre Curie.

BERNOULLI, DANIEL (1700–1782), SWITZERLAND

Bernoulli is known for his wide variety of work in mathematics, hydrodynamics, vibrating systems, probability, and statistics. Bernoulli's law of fluid dynamics states that the total energy of fluid pressure, gravitational potential energy, and kinetic energy of a moving fluid remains constant. For liquid flowing in a pipe, an increase in velocity occurs simultaneously with decrease in pressure. Bernoulli's formula has numerous practical applications in the field of aerodynamics, where it is considered in the study of flow over airfoils—such as wings, propeller blades, and rudders. Bernoulli's law is used when designing a venturi throat—a constricted region in the air passage of a carburetor that causes a reduction in pressure, which in turn causes fuel vapor to be drawn out of the carburetor bowl. Today, we write Bernoulli's law as:

$$\frac{v^2}{2} + gz + \frac{p}{\rho} = C$$

For example, the pressure of many fluids changes within a pipe according to this formula. Here, v is the fluid velocity, g the acceleration due to gravity, z the elevation (height) of a point in the fluid, p the pressure, and ρ the fluid density; C is a constant.

BOHR, NIELS (1885–1962), DENMARK

Bohr made important contributions to our understanding of atomic structure and quantum theory. The Bohr model of the atom included discrete energy levels of electrons. Electrons may jump from one energy level to another. His complementarity principle is a fundamental principle of quantum mechanics.

BOLTZMANN, LUDWIG (1844–1906), AUSTRIA

Boltzmann is famous for his interpretation of entropy as a measure of the disorder of a system due to thermal motion of molecules of the system. Thus, if the temperature is low, adding a quantity of heat to the system may elicit a relatively large additional disorder in the thermal motion of molecules in the system. He formulated a mathematical relationship between entropy S and molecular motion. This relationship is expressed as $S = k \cdot \log(W)$, where W is the number of possible states of the system, and k is Boltzmann's constant that gives S in useful units. This equation for S is engraved on Boltzmann's tombstone in Vienna.

BOYLE, ROBERT (1627–1691), IRELAND

Boyle's law showed the inverse relationship between the pressure P and the volume V of a gas in a container held at a constant temperature. Boyle observed that the product of the pressure and the volume is nearly constant.

DE BROGLIE, LOUIS (1892–1987), FRANCE

De Broglie is known for his postulates on the wave nature of electrons, and he suggested that all matter has wave properties with associated wavelengths. His equation $\lambda = h/mv$ shows the relations of the mass (m), velocity (v), and wavelength (λ) of a wave-particle; h is Planck's constant.

CARNOT, NICOLAS LÉONARD SADI (1796–1832), FRANCE

Carnot is sometimes referred to as the father of thermodynamics, which is the branch of physics concerned with the interrelationships of heat, temperature, energy, and work. In his 1824 monograph *Reflections on the Motive Power of Fire*, Carnot discussed his influential theory on the efficiency of heat engines.

CAVENDISH, HENRY (1731–1810), ENGLAND

Cavendish is famous for his experiment to "weigh the earth" (i.e., determine earth's density) and for his unpublished work on the dependence of the force between charged objects on distance and charge.

CHANDRASEKHAR, SUBRAHMANYAN (1910–1995), INDIA, UNITED STATES

Chandrasekhar is famous for his theories related to black holes and other massive stars. He also focused on stellar structure and dynamics, the theory of white dwarfs, general relativity, and gravitational wave theory.

CLAUSIUS, RUDOLF (1822–1888), GERMANY

Clausius contributed to the science of thermodynamics. Clausius's law, which is the second law of thermodynamics in one of its early formulations, states that the total entropy, or disorder, of an isolated system tends to increase over time as it approaches a maximum value. For a closed thermodynamic system, entropy can be thought of as a measure of the amount of thermal energy unavailable to do work. The entropy change of the universe, for any given process, must be greater than or equal to zero. Heat flows spontaneously from a hotter to a colder object but not vice versa.

COPERNICUS, NICOLAUS (1473–1543), ROYAL PRUSSIA, KINGDOM OF POLAND

Copernicus postulated a heliocentric model of the universe that placed the sun, instead of the earth, at the center. He is famous for *De revolutionibus orbium coelestium* (*On the Revolutions of the Celestial Spheres*), published shortly before his death in 1543.

DE COULOMB, CHARLES-AUGUSTIN (1736–1806), FRANCE

Coulomb is best known for his law that states that the force of attraction or repulsion between two electric charges is proportional to the magnitude of the charges and inversely proportional to the square of their separation distance. More particularly, the magnitude of the force F between two point charges in free space is given by

$$F = \frac{1}{4\pi\varepsilon_0} \frac{q_1 q_2}{r^2}$$

where q_1 and q_2 are the magnitudes of the charges, r is the distance between the charges, and ε_0 is the permittivity of free space. If the charges have the same sign, the force is repulsive. If the charges have opposite signs, the force is attractive.

DIRAC, PAUL (1902–1984), ENGLAND

Dirac is known for the Dirac equation, which can be written as:

$$\left(\alpha_0 mc^2 + \sum_{j=1}^{3} \alpha_j p_j c\right)\psi(\mathbf{x},t) = i\hbar\frac{\partial \psi}{\partial t}(\mathbf{x},t)$$

The equation describes electrons and other elementary particles in a way that is useful in both quantum mechanics and the special theory of relativity. In this equation, m is the rest mass of the electron, $\hbar$ is the reduced Planck's constant (1.054×10^{-34} J·s), c is the speed of light, p is the momentum operator, $\mathbf{x}$ and t are the space and time coordinates, and $\psi(\mathbf{x},t)$ is a wave function; α is a linear operator that acts on the wave function. The equation predicted the existence of antiparticles and in some sense "foretold" their experimental discovery. The discovery of the positron, the antiparticle of the electron, was a fine example of the usefulness of mathematics in modern theoretical physics. More generally, Dirac is famous for his important contributions to the development of both quantum mechanics and quantum electrodynamics.

DOPPLER, CHRISTIAN (1803–1853), AUSTRIA

Doppler is famous for the Doppler effect, his principle that the observed frequency of a wave depends on the relative motion of the source and the observer (or detector). He also suggested that the observed color of light from a star should change as a result of the star's velocity relative to earth.

EINSTEIN, ALBERT (1879–1955), GERMANY

Einstein is acclaimed for his contributions to the special and general theories of relativity, the mass-energy equivalence formula ($E = mc^2$), the photoelectric effect, and the theory of Brownian motion. In a poll conducted by *Physics World* in 1999, Albert Einstein was voted the greatest physicist of all time, pushing Isaac Newton into second place. Physicist Brian Greene, who participated in the poll, said, "Einstein's special and general theories of relativity completely overturned previous conceptions of a universal, immutable space, and time, and replaced them with a startling new framework in which space and time are fluid and malleable."

FARADAY, MICHAEL (1791–1867), ENGLAND

Faraday contributed to the fields of electromagnetism and electrochemistry. His induction law states that a changing magnetic field produces an electric field. James Clerk Maxwell expressed the change in magnetic flux and its relation to the induced electromotive force (ε or emf) in what we call Faraday's law of induction: $\varepsilon = -d\phi_m/dt$. Here, ϕ_m is flux of the magnetic field through a circuit.

FERMI, ENRICO (1901–1954), ITALY

Fermi is known for his work on the world's first artificial nuclear reactor (at the University of Chicago)—as well as for his research involving quantum theory, nuclear chain reactions, and other areas of particle physics.

FEYNMAN, RICHARD (1918–1988), UNITED STATES

Feynman developed the path integral formulation of quantum mechanics, the theory of quantum electrodynamics (a theory of the interaction between matter and light), and Feynman diagrams, while contributing to numerous other areas in physics.

FOURIER, JOSEPH (1768–1830), FRANCE

Fourier applied what we now call Fourier series to studies of heat transfer and vibrations. Fourier's law of heat conduction states that the rate of heat flow between two points in a material is proportional to the difference in the temperatures of the points and inversely proportional to the distance between the two points.

FRAUNHOFER, JOSEPH (1787–1826), GERMANY (BAVARIA)

Fraunhofer discovered dark absorption lines (Fraunhofer lines) in the sun's spectrum. It was later determined that these lines are caused by photon absorption by chemical elements in the solar atmosphere.

GALILEI, GALILEO (1564–1642), ITALY

Galileo is famous for his telescope improvements and astronomical observations that supported the heliocentric model of the solar system. He is well known for his observations of the four largest moons of Jupiter, analysis of sunspots, and experiments with falling bodies.

GAUSS, CARL FRIEDRICH (1777–1855) GERMANY

Gauss contributed to various fields of mathematics, along with electrostatics, astronomy, and optics. Gauss's laws demonstrated that the electric flux across any closed surface is proportional to the net electric charge enclosed by the surface. The magnetic flux across any closed surface is zero. More particularly, Gauss's law for electricity provides the relationship between the electric flux Φ flowing out of a closed surface and the electric charge enclosed by the surface:

$$\Phi = \oint_S \mathbf{E} \cdot d\mathbf{A} = \frac{1}{\varepsilon_o} \int_V \rho \cdot dV = \frac{q_A}{\varepsilon_o}$$

Here, $\mathbf{E}$ is the electric field. Gauss's law for magnetism is one of the fundamental equations of electromagnetism and is a formal way of stating the conclusion that no isolated magnetic poles exist. The law may be represented in equation form as:

$$\Phi_B = \oint_S \mathbf{B} \cdot d\mathbf{A} = 0$$

The law indicates that the net magnetic flux Φ_B across any closed surface is zero. $\mathbf{B}$ is the magnetic field.

GELL-MANN, MURRAY (B. 1929), UNITED STATES

Gell-Mann coined the term "quark" and is famous for contributions related to the theory and classification of elementary particles, effective complexity, and strangeness (a property of particles, expressed as a quantum number, for describing the decay of particles).

GIBBS, JOSIAH WILLARD (1839–1903), UNITED STATES

Gibbs is known for his work on the applications of thermodynamics. Also, together with Ludwig Boltzmann and James Clerk Maxwell, he helped create the field of statistical mechanics, which explains the laws of thermodynamics in terms of the statistical properties of collections of particles. Gibbs also worked on the application of Maxwell's equations in the area of optics.

HAWKING, STEPHEN (B. 1942), ENGLAND

Hawking is famous for his collaboration with Roger Penrose on gravitational singularity theorems, for his prediction that black holes emit radiation, and for theories of cosmology that involve the general theory of relativity and quantum mechanics. He is also a well-known science popularizer and author of the best seller *A Brief History of Time*, published in 1988.

HEISENBERG, WERNER (1901–1976), GERMANY

Heisenberg posited that the position and the velocity of an object cannot both be known with high precision at the same time. Specifically, the more precise the measurement of position, the more imprecise the measurement of momentum, and vice versa. The most common expression of this principle depicts the relations between the position x and the momentum p of a particle in space:

$$\Delta x \Delta p \geq \frac{\hbar}{2}$$

VON HELMHOLTZ, HERMANN LUDWIG FERDINAND (1821–1894), GERMANY

von Helmholtz is known for his contributions to theories on the conservation of energy and for his work in electrodynamics and thermodynamics. In the 1850s, he created a device (the Helmholtz resonator), which he used to study and identify pitches or frequencies in music and other sounds.

HERTZ, HEINRICH (1857–1894), GERMANY

Using electrical instruments, Hertz proved the existence of electromagnetic waves, which were suggested by James Clerk Maxwell's electromagnetic theory of light. Hertz showed that light and heat are electromagnetic radiation.

HOOKE, ROBERT (1635–1703), ENGLAND

Hooke's law of elasticity ($F = -kx$) states that if an object, such as a metal rod or spring, is elongated by some distance x, the restoring force F exerted by the object is proportional to x. Hooke inspired the increasing use of microscopes for scientific exploration. He also postulated the inverse-square law of gravity, but lacked the mathematical know-how to prove it. Although Hooke did not discover the universal law of gravitation, it appears that he did contribute to Newton's thinking on the subject.

HUBBLE, EDWIN (1889–1953), UNITED STATES

Hubble is famous for showing that the greater the distance a galaxy (or galactic cluster) is from an observer on earth, the faster the galaxy recedes. The distances between galaxies are continuously increasing; therefore, the universe is expanding. (Belgian priest and astronomer Georges Lemaître also proposed the theory of the expansion of the universe.) Additionally, Hubble showed that the universe extends beyond the Milky Way galaxy.

HUYGENS, CHRISTIAAN (1629–1695), DUTCH REPUBLIC

Huygens is acclaimed for his telescopic studies of the rings of Saturn, the discovery of Saturn's moon Titan, and the invention of the pendulum clock. He founded the wave theory of light and contributed to the science of dynamics.

JOULE, JAMES PRESCOTT (1818–1889), ENGLAND
Joule's law of electric heating states that the amount of heat produced by a steady electric current through a conductor is proportional to the resistance of the conductor, to the square of the current, and to the duration of the current. Joule confirmed that the various forms of energy—mechanical, electrical, and heat—are essentially the same and can be converted to one another. This led to the law of conservation of energy and the first law of thermodynamics.

KEPLER, JOHANNES (1571–1630), GERMANY
Kepler is famous for his three laws that describe the motions of planets about the sun. One of the laws states that all the planets in our solar system move in elliptical orbits, with the sun at one focus of the ellipse.

LEMAÎTRE, GEORGES (1894–1966), BELGIUM
This priest and astronomer proposed a theory of the expansion of the universe (see entry on Edwin Hubble). He also proposed a "primeval atom" or "cosmic egg" hypothesis for the origin of the universe that is related to the modern Big Bang theory.

LORENTZ, HENDRIK (1853–1928), NETHERLANDS
Lorentz contributed to the explanation of the Zeeman effect (the splitting of spectral lines). He also developed mathematical transformations used by Albert Einstein to describe space and time. These formulas describe the increase of mass, decrease in length, and dilation of time for moving bodies.

MACH, ERNST (1838–1916), AUSTRIA
Mach contributed to various principles of optics, mechanics, and shock waves, and other aspects of wave dynamics. Mach's principle suggests that inertia results from a relationship of an object with all the rest of the matter in the universe, an idea that influenced Einstein and his theories of relativity.

MAXWELL, JAMES CLERK (1831–1879), SCOTLAND
Maxwell formulated a set of elegant and concise equations that describe electricity, magnetism, and optics as manifestations of the electromagnetic field. Maxwell demonstrated that electric and magnetic fields travel through space as waves moving at the speed of light.

MEITNER, LISE (1878– 1968), AUSTRIA, SWEDEN
Meitner was a member of the team that discovered nuclear fission, in which the nucleus of an atom splits into smaller (lighter) nuclei. Element 109, meitnerium, is named in her honor.

MILLIKAN, ROBERT (1868–1953), UNITED STATES
Millikan is known for his experiments to determine the electric charge carried by a single electron and also for his experiments to verify an equation introduced by Albert Einstein to describe the photoelectric effect. He also proved that a certain form of radiation came from beyond earth; he named this radiation "cosmic rays."

NEWTON, ISAAC (1642–1727), ENGLAND
Newton is famous for his works relating to classical mechanics, optics, calculus, power series, the binomial theorem with non-integer exponents, the law of universal gravitation, and Newton's method for approximating the roots of a function: $x_{n+1} = x_n - f(x_n)/f'(x_n)$. Newton's second law of motion can be written as $\mathbf{F} = d\mathbf{p}/dt$, where $\mathbf{p}$ is the momentum, $\mathbf{F}$ is the applied force, and $d\mathbf{p}/dt$ is the rate of change in momentum. Newton's law of universal gravitation for two point masses ($F = Gm_1m_2/r^2$) is for idealized bodies whose size is very small compared to separation distances. Here, F is the magnitude of the gravitational force between the two masses, G is the gravitational constant, m_1 is the mass of one of the masses, m_2 is the mass of the other point mass, and r is the distance between the masses. The value of G is usually given as 6.67×10^{-11} Nm²·kg⁻².

Newton's law of cooling states that the rate of heat loss of a body is proportional to the difference in temperatures between the body and its surroundings. Newton was also famous for his studies on the composition of white light and his pioneering efforts in infinitesimal calculus. His most famous work was *Philosophiae Naturalis Principia Mathematica* (*The Mathematical Principles of Natural Philosophy*, 1687).

OHM, GEORG (1789–1854), GERMANY
Ohm's law states that the current flow through a conductor is proportional to the voltage and inversely proportional to the resistance.

ØRSTED, HANS (1777–1851), DENMARK
Ørsted discovered that electric currents (e.g., in a wire) create magnetic fields, an important step in the development of the study of electromagnetism.

PAULI, WOLFGANG (1900–1958), AUSTRIA
The Pauli exclusion principle, in essence, states that no pair of identical particles can simultaneously occupy the same quantum state. For example, two electrons occupying the same atomic orbital must have opposite spins. This principle plays a fundamental role in quantum theory and applies to fermions (such as electrons, protons, and neutrons) but not to bosons (such as photons).

PLANCK, MAX (1858–1947), GERMANY
Planck was the originator of quantum theory. His explanation of blackbody radiation suggests that electromagnetic energy can be emitted only in quantized forms. Planck's formulation is notable because it incorporates the earliest known practical application of quantum theory.

RÖNTGEN, WILHELM (1845–1923), GERMANY
Röntgen is famous for the discovery of X-rays, which earned him the first Nobel Prize in Physics in 1901.

RUTHERFORD, ERNEST (1871–1937), ENGLAND

Rutherford is considered the father of nuclear physics. He discovered the concept of radioactive half-life and proved that radioactivity involves the transmutation of one element to another. The Rutherford model of the atom suggests that a central atomic nucleus is concentrated into a very small volume in comparison to the rest of the atom. Most of the atom's mass resides in the nucleus.

SALAM, MOHAMMAD ABDUS (1926–1996), PAKISTAN

Salam is known for his work in nuclear physics, and, in particular, for helping to formulate the electroweak theory, which explains the unity of electromagnetism and the weak nuclear force. He also worked on supersymmetry and contributed to modern theories on neutrinos, neutron stars, and black holes.

SCHRÖDINGER, ERWIN (1887–1961), AUSTRIA

Schrödinger contributed to quantum theory and wave mechanics. Schrödinger's wave equation describes ultimate reality in terms of wave functions and probabilities:

$$i\hbar\frac{\partial}{\partial t}\Psi(\mathbf{r},t) = \left[\frac{-\hbar^2}{2m}\nabla^2 + V(\mathbf{r},t)\right]\Psi(\mathbf{r},t)$$

Here, ψ is the wave function, m is the particle's mass, and V is its potential energy.

SKŁODOWSKA-CURIE, MARIE (1867–1934), POLAND, FRANCE

Curie conducted pioneering research on radioactivity and was the first woman to win a Nobel Prize. She discovered two elements, polonium and radium.

THOMSON, JOSEPH JOHN ("J. J.") (1856–1940), ENGLAND

Thomson showed that cathode rays are composed of negatively charged particles (electrons), a finding that revolutionized our knowledge of atomic structure.

THOMSON, WILLIAM (LORD KELVIN) (1824–1907), IRELAND, SCOTLAND

Thomson contributed to the mathematical analysis of electricity and formulation of the first and second laws of thermodynamics. He is also known for providing a value for absolute zero. Absolute zero corresponds to -273.15 °C or -459.67 °F.

TORRICELLI, EVANGELISTA (1608–1647), ITALY

Torricelli is famous for his invention of the barometer and for Torricelli's law in fluid dynamics, relating the speed of a fluid flowing out of an opening (e.g., of a container) to the height of the fluid above the opening. A torricellian vacuum can be created by filling a glass cylinder (closed at one end) with mercury and then inverting the cylinder into a bowl to contain the mercury.

TOWNES, CHARLES HARD (B. 1915), UNITED STATES

Townes is famous for his work on the theory and application of the maser (an acronym for microwave amplification by stimulated emission of radiation), as well as for his work in quantum electronics associated with both maser and laser devices. Today, lasers have many important applications in medicine, communications, and information-processing technologies.

VOLTA, ALESSANDRO (1745–1827), DUCHY OF MILAN, ITALY

Volta is acclaimed for his invention of the battery, a source of continuous electric current.

WEBER, WILHELM (1804–1891), GERMANY

Weber is famous for his work on the earliest electromagnetic telegraphs as well as his investigations of terrestrial magnetism.

WHEELER, JOHN (1911–2008), UNITED STATES

Wheeler is known for his contributions and ideas related to nuclear fission, the Breit-Wheeler process, geometrodynamics, general relativity, gravitation, quantum mechanics, unified field theory, Wheeler-Feynman absorber theory, and Wheeler's delayed choice experiment. He also popularized the terms "quantum foam," "wormhole," and "black hole."

YOUNG, THOMAS (1773–1829), ENGLAND

Young helped establish the wave theory of light. With his double-slit experiment, he demonstrated interference in the context of light as a wave. When Young passed light through two nearby pinholes, he showed that the light beams produced a series of stripes of light.

SELECTED BIBLIOGRAPHY

Arons, Arnold. *Development of Concepts of Physics*. Reading, MA: Addison-Wesley, 1965.

Atiyah, Michael. "Pulling the Strings." *Nature*, vol. 438, no. 7071 (2005): 1081–82.

Brockman, John. *What We Believe but Cannot Prove*. New York: Harper Perennial, 2006.

Bryson, Bill. *A Short History of Nearly Everything*. New York: Broadway Books, 2004.

Crease, Robert. "The Greatest Equations Ever." *Physics World*, vol. 17, no. 10 (2004): 19–23.

Cropper, William. *Great Physicists*. New York: Oxford University Press, 2004.

Davies, Paul. *Superforce*. New York: Simon & Schuster, 1984.

Dirac, Paul. "The Relation between Mathematics and Physics." *Proceedings of the Royal Society (Edinburgh)*, vol. 59 (1938–1939): 122–129.

Farmelo, Graham. *It Must Be Beautiful*. London: Granta Books, 2003.

Feynman, Richard. *The Feynman Lectures on Physics*. Boston: Addison-Wesley, 1964.

Feynman, Richard. *The Character of Physical Law*. London: BBC, 1965.

Frayn, Michael. *The Human Touch*. New York: Metropolitan Books, 2007.

Gardner, Martin. "Order and Surprise." *Philosophy of Science*, vol. 17, no. 1 (1950): 109–117.

Guillen, Michael. *Five Equations That Changed the World*. New York: Little, Brown, 1995.

Halliday, David, and Robert Resnick. *Physics*. Hoboken, N.J.: Wiley, 1966.

Hawking, Stephen. *Black Holes and Baby Universes and Other Essays*. New York: Bantam, 1993.

Icke, Vincent. *The Force of Symmetry*. New York: Cambridge University Press, 1995.

Johnson, George. "Why Is Fundamental Physics So Messy?" in Hodgeman, John, ed., "What We Don't Know." *WIRED*, vol. 15, no. 2 (2007): 104–124.

Krauss, Lawrence. *Fear of Physics*. New York: Basic Books, 1993.

Krauss, Lawrence. *The Physics of Star Trek*. New York: Basic Books, 1995.

Manin, Yuri. "Mathematical Knowledge: Internal, Social, and Cultural Aspects." In *Mathematics as Metaphor*. Providence, R.I.: American Mathematical Society, 2007.

Pesic, Peter. "The Bell & the Buzzer: On the Meaning of Science." *Daedalus*, vol. 132, no. 4 (2003): 33–44.

Pickover, Clifford. *Archimedes to Hawking*. New York: Oxford University Press, 2008.

Pickover, Clifford. *The Physics Book*. New York: Sterling, 2011.

Schwartz, Joseph. *The Creative Moment*. New York: HarperCollins, 1992.

Smolin, Lee. "Never Say Always." *New Scientist*, vol. 191, no. 2570 (2006): 30–35.

Trefil, James. *The Nature of Science*. New York: Houghton Mifflin Harcourt, 2003.

Zebrowski, George. "Time Is Nothing but A Clock." *OMNI*, vol. 17, no. 1 (1994): 80–82, 114–115.

IMAGE CREDITS

NASA: N.A.Sharp, NOAO/NSO/Kitt Peak FTS/AURA/NSF: 69; NASA/Swift/Mary Pat Hrybyk-Keith and John Jones: 254; 104; 212

© Clifford A. Pickover: 334

Shutterstock: © -baltik-: 200; © Alison Achauer: 192; © agsandrew: 15, 17, 19, 26, 40, 50, 51, 78, 93, 100, 108, 110, 111, 117, 120, 142, 146, 170, 172, 174, 177, 178, 187, 191, 224, 228, 240, 241, 256, 258, 273, 274, 282, 290, 291, 302, 307, 308, 310–312, 317, 319, 322, 323, 326, 332, 363; © Alegria: 279; © AlexussK: 14; © Algol: 28, 169; © alicedaniel: 261; © alri: 173; © Manuel Alvarez: 6; © Anenome: 24; © Anikakodydkova: 204; © Anneka: 198; © Matt Antonino: 87; © argo: 182; © Vladimir Arndt: 83; © Rachael Arnott: 107; © ArTDi101: 158; © Ase: 145; © asharkyu: 260; © Stefan Ataman: 265; © Zvonimir Atletic: 344; © Awe Inspiring Images: 294; © awstock: 233; © B Calkins: 205; © Anton Balazh: 31; © Denis Barbulat: 119; © Marcio Jose Bastos Silva: 139; © Fernando Batista: 324; © Martin Bech: 44; © ben44: 195; © Tomasz Bidermann: 53; © bluecrayola: 89, 112; © BlueRingMedia: 353; © Boris15: 330; © Simon Bratt: 238; © Brocreative: 295; © Michael D Brown: 33; © Linda Bucklin: 297; © Boris Bulychev: 70; © bumihills: 77; © Martin Capek: 331; © Vladimir Caplinskij: 48; © carlos castilla: 252; © Catmando: 202; © Champiofoto: 84; © chaoss: 369; © Ivan Cholakov Gostock-dot-net: 94; © clawan: 168; © Steve Collender: 210; © Color Symphony: 102; © Perry Correll: 122; © cosma: 160; © jamie cross: 292; © danielo: 301; © Andrea Danti: 11; © Dashu: 219; © Elaine Davis: 12; © dedek: 194; © denniz: 43; © Digital Saint: 76; © diversepixel: 18; © Dja65: 276; © Dolgopolov: 235, 357; © Dragonfly22: 82, 114, 167, 189; © edobric: 346; © ErickN: 29; © EtiAmmos: 99; © Everett Collection: 75, 103, 132, 157, 216, 270; © F. ENOT: 352; © Paul Fleet: 354; © Fotocostic: 115, 300; © Markus Gann: 206; © Christos Georgiou: 365; © Stefano Ginella: 268; © Jeff A. Goldberg: 109; © Oleg Golovnev: 163; © Dmitri Gomon: 298; © Jennifer Gottschalk: 133, 165, 315, 338, 355; © GrAl: 135; © Gwoeii: 86, 336; © Jorg Hackemann: 219; © handy: 349; © Angela Harburn: 140; © patrick hoff: 357; © Christopher S. Howeth: 74; © Igor Zh.: 85; © ikonstudios: 81; © imagedb.com: 52; © In-Finity: 362; © Eugene Ivanov: 47, 60, 335; © Jaswe: 30, 46, 159, 208; © Jezper: 257; © Petr Jilek: 223; © Kjersti Joergensen: 207; © JPRichard: 221; © justasc: 213, 278; © Kaetana: 105; © Victoria Kalinina: 343; © Panos Karas: 236; © Kasza: 101; © katalinks: 196; © KeilaNeokow EliVokounova: 36, 144, 161, 185, 217, 287; © Brian Kinney: 327; © Richard Koczur: 149; © koya979: 150; © Max Krasnov: